ROUTLEDGE LIBRARY EDITIONS:
JAPAN

JAPANESE WHALING

JAPANESE WHALING

End of an Era?

ARNE KALLAND & BRIAN MOERAN

Volume 76

Routledge
Taylor & Francis Group

LONDON AND NEW YORK

First published 1992 by Routledge

This edition first published in 2011
by Routledge
2 Park Square, Milton Park, Abingdon, Oxon, OX14 4RN

Simultaneously published in the USA and Canada
by Routledge
52 Vanderbilt Avenue, New York, NY 10017

Routledge is an imprint of the Taylor & Francis Group, an informa business

British Library Cataloguing in Publication Data
A catalogue record for this book is available from the British Library

ISBN 13: 978-0-415-84544-1(pbk)
ISBN 13: 978-0-415-58819-5 (hbk)

Publisher's Note
The publisher has gone to great lengths to ensure the quality of this reprint but
points out that some imperfections in the original copies may be apparent.

Disclaimer
The publisher has made every effort to trace copyright holders and would
welcome correspondence from those they have been unable to trace.

Japanese Whaling

End of an Era?

Arne Kalland & Brian Moeran

CURZON
PRESS

Scandinavian Institute of Asian Studies
Monograph Series No. 61

First published 1992 by
Curzon Press Ltd.
St John's Studios
Church Road
Richmond
Surrey TW9 2QA

Reprinted 1993

ISBN 0 7007 0244 X

British Library Cataloguing in Publication Data
A CIP catalogue record for this book
is available from the British Library

Printed in England by Athenaeum Press, Newcastle

Table of Contents

Acknowledgements

This book could not have been completed without the generous assistance of the following institutions and organizations: in Japan, the Institute of Cetacean Research, the Japan Whaling Association, and the Small Type Whaling Association; in Denmark, the Nordic Institute of Asian Studies; and in Great Britain, the School of Oriental and African Studies, University of London. Numerous individuals in one way or another involved in whaling also patiently answered our many questions and explained to us all kinds of historical, social and cultural details connected with Japanese whaling. For a number of reasons, it would be unwise perhaps to single out these individuals here, but we would like to extend special thanks to certain members of the town administrations of Arikawa, Oshika and Ukushima, as well as to their Fishing Cooperative Associations and shrine and temple officials, for their kindness and generosity during our fieldwork visits.

Others who have assisted us in numerous and varied ways include Misaki Shigeko, Takahashi Junichi, and Ted Bestor. The last two, together with the editors of *MAST*, we would like to thank for permission to reprint (as Chapter 5) a substantial part of a paper that we published together in that journal. We are also grateful to our colleagues, particularly to Milton Freeman and Iwasaki Masami, who participated in the International Workshop on Small Type Coastal Whaling in Japan in April 1988, and who both at that time and thereafter have freely discussed the findings of their research with us.

Finally, our thanks go to the following for permission to use their photographs: Institute of Cetacean Research (*Plates* 4, 6, 7, 8, 9, 13); Sasaki Hideo (*Plates* 10, 11, 12, 26); Suntory Beer Company (*Plate* 27); Taiji Town Office (*Plate* 24); Toba Yōjirō (*Plates* 20, 22); and Tsuboi Naokatsu (*Plate* 23). All other photographs used in this book were taken by the authors.

Taiji, January 1992

CHAPTER 1

Whales, Whaling, and Japan

The aim of this book is to give a social anthropological account of
whaling in Japan. In so doing, we present the first comprehensive
account in English of the history of Japanese whaling, showing how
it has given rise to a particular kind of culture, and we discuss what
happens when that culture is threatened. At the same time, we explain
the work organization of those employed in whaling; the role of
whaling companies in local and national economies; and the role of
the whale in the establishment and maintenance of local community
identity (ritual, food, gift giving). In short, we are concerned with
what we term a "whaling culture".

Given the antipathy most westerners now feel towards the issue
of whaling, it may be asked why we have chosen to write such a
book. In the forum of the International Whaling Commission's annual
meetings, the Japanese Government has, over the years, tried to argue
for the continuation of whaling from three different points of view.
Firstly, it has suggested that there is no *ecological* reason to abandon
commercial whaling entirely, since scientific reports show that there
are adequately sustainable stocks of certain species of whale.
Secondly, it has simultaneously argued along *economic* grounds that
whaling should be continued since those involved otherwise suffer
undue hardship. Finally, in recent years, the Japanese delegation has
focussed on *cultural* arguments and petitioned for small type coastal
whaling (STCW)[1] to be considered along the same lines as aboriginal
whaling, accepted as a separate category by members of the International Whaling Commission (IWC) in 1981.

It was in the context of the last line of reasoning that a group of

---

[1] i.e. hunting minke whales and small toothed-whales in Japanese coastal waters,
using boats of less than 48 tons.

a dozen anthropologists from six different countries (and including this book's two authors) was called together in April 1988 for a workshop on small type coastal whaling. Their brief was to investigate the social, economic and cultural significance of STCW in those regions of Japan where this form of whaling used to be practiced prior to the moratorium on all forms of commercial whaling. They were thus being asked to provide factual information that might help decision-makers resolve the question of whether STCW could in fact legitimately be regarded as similar in certain significant features to the kind of small-scale whaling practised by some aboriginal societies.

We ourselves, however, wished to extend the discussion to Japanese whaling in general, in order to develop further the key concept of "whaling culture", which in the 1988 report was defined as "the shared knowledge of whaling transmitted across generations" (Akimichi et al. 1988:75). This shared knowledge was seen as consisting of "a number of different socio-cultural inputs: a common heritage and world view, an understanding of ecological (including spiritual) and technological relations between human beings and whales, special distribution processes, and a food culture" (*ibid.*). We therefore embarked upon further periods of fieldwork in some of Japan's whaling communities and it is the results of this research that are published here.

It should be stressed that in a number of respects this work has been by no means easy. In the first place, the issue of whales and whaling is provocative. In modern industrial countries like the United States or Great Britain, large numbers of people feel that they have a right to speak out against whaling and to condemn those whom they see as being involved in "barbaric" activities. At least one of the original members attending the International Workshop in 1988 has since been told that s/he would be better advised not to get involved further in research on whaling, if s/he places any value on an academic reputation and does not want to become a scapegoat for anti-Japanese feelings. At a somewhat less threatening level, the authors themselves received a letter from one well known publishing company in which a senior editor exclaimed "I hate this vile industry—and I am sure you do, too", before continuing to suggest certain ways in which he felt that this book might be "improved" and so become more acceptable to members of a leading environmental organization.

Such difficulties bring us to the crux of the issue of whaling. We would like to make it plain that, in our opinion, whaling, as it is at present discussed in international meetings and by the various media, is to a large extent not an ecological, but a moral, issue and that many of the various attempts to protect whales are based more on ethico-political reasoning than on strictly environmental issues. This is now even admitted by certain animal rights groups which recognize that "at least some species of whales could easily sustain a resumed, strictly regulated harvest, without threatening species survival. Therefore, now it is an appropriate time to face and discuss *the moral and ethical issues* involved in the commercial harvesting of whales" (Barstow 1989:10, italics added). Similar sentiments have been voiced by, for example, the U.S. Commissioner to the International Whaling Commission (IWC), John Knauss, and the British Minister of Agriculture and Fisheries, John Gummer.[2]

Our first question with regard to the issue of whaling is: why do certain groups of people choose certain forms of wild life as the object of their attention? Obviously, there are differences among the various environmental and animal rights groups active in the world today.[3] Some keep strictly to environmental issues and accept the hunting of any species, provided that stocks are sufficient to sustain the harvest thereof, while simultaneously working for the total protection of those species threatened with extinction. We ourselves are in full sympathy with this view.

On the other hand, there are those animal rights groups which condemn *any* taking of life, regardless of whether there are or are not sufficient numbers of living creatures to make their killing acceptable on environmental grounds. This may on the face of it seem to be a consistent viewpoint (provided, of course, that its adherent is a

---

[2] *Marine Mammal News* (May 1991) and *The Times* (29 May 1991), respectively.

[3] By an "environmental group" we here mean a group of persons ("conservation-ists") who are concerned with environments as *systems* and who therefore work to secure habitats and species diversity and not for the well-being of individual animals (Lynge 1990:47). The latter are the concern of animal rights groups ("protectionists") which are against killing animals *per se*. There is no sharp line between these two types of groups, however (Wenzel 1991:36). Animal rights groups have become increasingly concerned with ecological systems, and environmental groups have recently engaged themselves in the protection of non-endangered species for—as we will presently show—other than ecological reasons. The confusion also afflicts U.S. lawmakers (Manning 1989).

vegetarian), but—as any Southeast Asian Buddhist monk will be the first to admit—well nigh impossible to carry out. Human beings, including vegetarians, necessarily impinge upon animal life. We kill germs and bugs; we use insecticides to grow our crops; by merely removing the original land cover in order to develop agriculture, we deprive countless animals and birds of their natural habitat. In short, life depends upon taking life (Schweitzer 1950:189).

Finally, there are those who—in our opinion—confuse environmental and ethical issues by arguing that it is morally wrong to kill certain mammals, such as whales or seals, while it is apparently acceptable to kill such others as cattle or moose. Here, a somewhat devious mode of reasoning is employed since those concerned often use environmental arguments to defend a species which is in no way endangered. In their attempts to satisfy their supporters, as well as obtain political support for their beliefs, they thus shift from the practical issue of *resource* management to a verbal rhetoric largely concerned with *information* management. Indeed, some groups appear to have chosen certain mammals in order to focus other people's attention on animal suffering, while at the same time opening the strings of their—at times bountiful—purses (Gulland 1988:45). In this sort of moral arena, there is little to rival the wide, "tearful" eyes of an "innocent" white "baby seal" waiting to be clubbed to death by a "heartless" hunter.[4]

Although it is not our purpose here to analyse fully the activities of environmental and animal rights groups,[5] a few comments are perhaps in order. After all, these days whales provide almost as effective a fund-raising image as seals have done (Herscovici 1985, Henke 1985). This leads us to ask: why *whales*? Why, for example, are whales better protected by U.S. law than any other animals, with the possible exception of other marine mammals, e.g. seals (Manning 1989:220)? Activists will immediately say that the answer to this question is obvious and simple: whales are endangered. We would ask

---

[4] Perhaps it should be pointed out here that the anthropomorphization of animals (as seen in the phrase "baby seal", rather than "seal pup") is very much a part of this information management (see Freeman 1990).

[5] Recently a number of articles critical of certain groups have appeared. Interested readers are advised to consult Eyerman and Jamison (1989); Fox (1991); Schwarz (1991); Spencer et al. (1991); and Taylor (1991). See also Hercegovici (1985) and Wenzel (1991).

whether this is true of all whales. It is a fact that some species of whale are (by some definitions of the term, at least) endangered, and that every effort should be made to protect them. However, many marine mammal stocks appear to be both healthy and abundant. According to the Director of the Northwest Alaska Fisheries Center, for example, at the end of the 1980s 57 species of cetacean were already at or near their original level of abundance (Aron 1988:104). These include minke, pilot, and Baird's beaked whales (the three species hunted by the Japanese STCW fleet), and Bryde's whale (previously hunted by the large type coastal whaling [LTCW] fleet until it was ordered to close down). The Antarctic stocks of minke whales, which are at present the focus of Japanese research whaling, are particularly abundant (Gulland 1988:44). The only species until recently hunted by coastal Japanese fleets and defined as "endangered" by the U.S. Endangered Species Act of 1973, is the sperm whale. Yet even the exploitable sperm whale population is reckoned to approach one million animals,[6] a figure which suggests that it cannot in a biological sense be perceived as an endangered species. This means that none of the species of whale taken by Japanese Antarctic, LTCW, and STCW fleets during the past decade has been endangered. In short, that a few species of whale are endangered does not in our opinion justify a total moratorium on all killing of all whales. After all, the fact that the Bengal tiger is seen to be endangered does not mean that we should protect all wild—or even tame—cats that still exist in sufficiently sustainable numbers.

So what is it about whales that attracts so much attention? For a start, the way in which Antarctic whaling used to be conducted—with numerous fleets ruthlessly exploiting the environment solely for profit—has led to a public outcry, headed by environmentalists, that all whales are on the verge of extinction. This has resulted in an image of the "poor" and "exploited" whale, which is then tacked on to a general mystique about whales that has evolved over the centuries in the West, and is to be found in literature from the *Bible* to *Moby Dick* and contemporary science fiction. This mystique is easily enhanced by the fact that whales form an anomalous category of

---

[6] The exploitable population of sperm whales is defined as consisting of male animals older than eleven years, and females older than ten years. This exploitable population is estimated to comprise between 40 and 60 per cent of the total sperm whale population (Aron 1988:104).

mammal (Douglas 1966). Like seals, they live in the sea rather than on land, and have fins like fish rather than legs like land mammals. Unlike fish, their tails—or flukes—are horizontal, rather than vertical. They also have no scales. Unlike seals and land mammals, but like men, they are not covered with heavy body hair. In short, whales are "betwixt and between" and hence singled out for special attention, allowing them in some ways to take on the characteristics of a "totem" for many environmentalists (Kalland 1991a).

At the same time, the fact that whaling has been carried out by a limited number of countries means that it is easy for those environmentalists involved in information management to pinpoint those whom they regard as responsible for the near extinction of some species of whales. This allows them then to focus on certain offenders, make them into scapegoats, and so transform a single disaster into a broad ecological issue. In short, it would seem as if in recent years whales (like seals) have become a metonym for the environment as a whole. As a leading anti-whaling campaigner, Sidney Holt (1985:12), has admitted: "Saving the whale is for millions of people a crucial test of their political ability to halt environmental destruction".

Such methods should not, of course, be taken to imply that anti-whalers are necessarily afflicted by some madness. But it does seem as if those who are so active in the protection of whales suffer from the kind of romanticism that is typical of those living in modern industrialized societies (see Williams 1973). This may, indeed, explain the IWC's agreement to allow a "primitive" people, the Eskimos, to hunt what has been regarded as one of the most endangered of all whales—the bowhead whale. "It's all right for them," goes the underlying, but unstated, line of reasoning. "They live in the wild and are close to nature." Certainly, this kind of attitude towards nature is a common characteristic of many urbanized peoples (including, let us hurriedly add, the Japanese), and we would do well to note that it is those living at least one step removed from the land or sea who tend to invest it with qualities that farmers or sailors, for example, would be among the last to accept.

In the midst of such romanticism, it is hardly surprising to find that "whale watching" is now becoming a trendy trade off the Californian, New England, Norwegian, and other coasts (including those of the Azores and Bonin Islands). Many people are attracted to ventures such as this, since it allows them to watch these splendid mammals "in their natural habitat". They, too, can "embrace life". Not only can they "save" whales, but they can "meet" them, "engage" them, even "touch" them perhaps. For them, as for the writer of the history of Greenpeace, there can be "a transcendent element lying at the center of the undertaking". They, too, can be said to be in search of their "Holy Grail" (Hunter 1979:150). Not only this, but "cetaceans increasingly appear to seek out and gain satisfaction from friendly encounters with humans, suggesting mutual enrichment as a common goal in human/cetacean relations in a modern world" (Barstow 1988:10). It would seem from all this, therefore, that not only can humans save whales, but that whales are actively involved in the planned salvation of urban man from boredom. Like certain parts of Africa, the ocean is fast becoming a wildlife preserve (witness the IWC's designation of the Indian Ocean as a "whale sanctuary"), where visitors can practise substitute hunting (shooting not animals or mammals, but film) and "experience" nature in all its "majesty". No doubt, in due course, people will be removed from the sea, in the same way that the Maasai have been turned off their land, in order to make way for the new "communion" between nature and man (Odner 1978:31-32).

In this respect, we should realize that those who suffer from the activities of anti-sealing and anti-whaling groups are easy targets for attack, since they live a long way from the centres of political power where decisions are made, and are in general extremely vulnerable. Some organizations seek issues which seem to be easy to win (Eyerman and Jamison 1989), and the international director of Greenpeace, Steve Sawyer, states that "our philosophy on issues is extraordinarily pragmatic. We choose the ones we fell we might be able to win" (Pearse 1991:40). One of the points that we make in this book is that, even though Japan may be a rich industrialized nation and hence generally able to fend for itself, its whalers are certainly not in a position to defend themselves against the attacks of those who object—at times violently—to the killing of whales.

If we then turn our attention to answering the related question of "why the sea?", we find that the romanticism noted above rears its head in a slightly different form. Both salt and water are important purifying agents which are used in religious rites throughout the world. The ocean, consisting largely of these two purifying agents, becomes *the* symbol of purity—of untouched nature—and thus stands in sharp contrast to the polluted soil on which we land mammals tread. Moreover, the sea is a good "issue" on which to focus because in general people do not know that much about what goes on in it; similarly, whales are also convenient because so little is known about them. To turn the problem the other way round, we might ask (along with some of our Japanese informants) what would happen if environmentalists were to decide to focus on the wholesale slaughter of cows, pigs or sheep as an "ecological" issue. Objections would be raised at once because, over the centuries, man (particularly western man) has built up a shared knowledge of the purpose of animal husbandry, of what it "means". After all, the counter-argument would run, animal husbandry consists of the kind of shared knowledge transmitted across generations, together with an understanding of ecological and technological relations between animals and man. But, in our opinion, this is precisely the characteristic of whaling culture as well. In other words, our argument is that we embrace—consciously or otherwise— an ideology about animal life which is for the most part lacking when it comes to marine life. By focussing on such issues as the slaughter of seals and whales, environmentalists have not had to combat an already existing set of ideals, but merely to create one. Our lack of knowledge about whales in general (even about the sheer variety of species of whale—over 70 in all—and the differences among them) has enabled protectionists to take the commonly believed characteristics of one species (the supposed intelligence of dolphins, for example)[7] and to apply them in blanket fashion to all whales regardless. By lumping together traits found in different species in this way, we soon arrive at an image of a "super whale"—a whale which is at the same time the largest animals on earth (true of the blue whale), has the largest brain (the sperm whale), has a large brain-to-body-weight ratio (the bottlenose dolphin), sings nicely (the humpback whale), is friendly (the gray whale), but endangered (the bowhead and the blue whales), and so on (Kalland 1991a).

Finally, the romanticism that we have here outlined is in many

---

[7] The intelligence of whales, and in particular of dolphins, is a matter of heated debate and relates primarily to the size and structure of the brain, as well as to certain mythical properties (like dolphins' ability to save human beings, to communicate with one another, and so on) (Prescott 1981; Pryor 1981). In fact, it would seem that in many respects the cetacean brain is not especially complex, and that its structure is quite primitive, for "even the most advanced cetacean brains seem to be stuck at a stage called the paralimbic-parinsular, which is the most primitive stage in land mammals ... such as hedgehogs and bats" (Klinowska 1988:46; see also Freeman 1990).

respects a critique of so-called "capitalist" culture in that the underlying presumption of most environmentalists appears to be that commercialism *per se* is bad and that whaling is only acceptable if carried out by aboriginal peoples as part of a subsistence economy.[8] This critique is, however, never carried to its logical conclusions. On the one hand, whale watching itself is now—in the United States alone—a multi-million dollar commercial enterprise (Barstow 1989:11), and more than 200 tour operators offer whale watching trips in North-America (Corringan 1991). That the value of a "non-consumptive use" of whales has surpassed the value of whaling itself is now—by a curious inversion of "logic"—being used as an argument to ban the latter (US Marine Mammal Commission 1991). On the other hand, most environmental and animal rights groups make extensive use of media coverage, thereby converting the objects of their attention into "events", and hence assimilating them to other "events" characterizing modern industrial societies—the Olympic Games, the Nobel Prize, or Live Aid—where the kind of fashion for saving whales that led to a pop concert being given on their behalf in the 1970s has now been expanded to include concerts for Ethiopians (Marcus 1989:277), Nelson Mandela and Amazonian tribes threatened by destruction of the tropical rain forests in which they live.

This predilection for media coverage also makes it necessary for such groups to be seen to score "victories". This means that, besides the singling out of weak or vulnerable groups for attack, the same criteria of success and efficiency used to evaluate the behaviour of politicians and big business in capitalist economies are also being applied by supporters of environmental and animal rights groups to their own campaigns. Thus, like politicians, environmentalists are dependent on a "fickle" electorate, so that they end up being very much part and parcel of the kind of social structure that they

---

[8] Incidentally, while the Eskimos are now allowed to eat the meat of the whales that they catch every year, the anti-commercial bias resulting in a world-wide ban on the trade of whale products prevents them from selling certain non-edible by-products, such as bowhead whale baleen. As a result, a traditional Japanese art form, the *bunraku* puppet theatre, faces difficulties since an entirely suitable substitute for this kind of baleen (used as part of the mechanism of the puppets themselves) does not appear to exist (personal communication, Nelson Graburn).

apparently set out to subvert.[9] At the same time, of course, by allowing such critiques to surface in a controlled manner, those in command of the capitalist economies also make them safe.

There are other contradictions in which these groups find themselves trapped. After all, when fighting crass commercialism as practised by big business and multinational enterprises, they end up making local people suffer by their actions (Herscovici 1985:23). Moreover, these local people living in remote areas very often end up by being dependent on either polluting or potentially polluting industries,[10] so that the net effect is to strengthen, rather than weaken, the capitalist system.

Which brings us to the question of funds. Let us here be clear about one thing: "environmentalism" is an industry with an enormous turnover and considerable profits to be made. Greenpeace, for example, has been reported to have an annual income of $36 million in Germany alone, and $157 million worldwide in 1990 (Spencer et al. 1991:174). One major problem facing any organization trying to pursue any moral cause is the financing of its campaign. Those pursuing environmentalist and animal rights causes are no exception to this rule and are obliged to seek funding from a variety of sources. Some groups receive large contributions from their many members,[11] but part of their funding inevitably comes from different sectors of the very system which they are criticizing. Thus it would seem that the anti-seal campaign has been strongly supported financially by someone intent on establishing a synthetic fur factory (Herscovici 1985:82). In the case of whaling, we might ask which industries or corporations were likely to benefit most from the collapse of the whale oil market and hence to provide funds to "environmental" and "animal rights" groups. Some Japanese informants—their suspicions aroused by the apparent lack of interest shown by certain environmental groups in a recent major Alaskan oil spill—have suggested that those pursuing the

---

[9] Since most politicians cannot afford to ignore the "green" vote in their respective countries, they automatically vote at such international gatherings as the IWC for issues that are seen by their constituents to be environmentally "safe".

[10] In this book we note, for example, how a nuclear plant has been established near the whaling community of Ayukawa, and oil installations near that of Arikawa.

[11] Accordning to *Forbes Magazine*, each of the 700,000 members of Greenpeace Germany pays $30 annual dues (Spencer et al. 1991:174).

anti-whaling campaign are in fact being financially supported by oil companies which wish to divert attention from their own activities and so avoid being accused of causing environmental damage. Such speculations are not pure fantasy. After all, the World Wildlife Fund (WWF) in Denmark has been sponsored by the Norwegian State Oil Company and has "sold", under its "adopt-a-whale" programme, a sperm whale sighted off the coast of northern Norway to a Danish chemical company. Apparently, certain organizations are able to sell a green image—or a green conscience—to individuals, companies, and countries which need it (Kalland 1991b). At the same time, we should not forget the fact that it is quite possible for environmental organizations to play different Japanese industries off against the adverse publicity caused by Japanese whaling in their search for funds.

This brings us—albeit in a circular manner, perhaps—to the management of whale resources and the role of the IWC. Most of the "commercial" whales are highly migratory and, as a result, are generally not limited to the waters of any one nation (however extended its coastal jurisdiction might be). This means that, in terms of resource management, whales are "common property". In other words, "no single user has exclusive use rights to the resource nor can he prevent others from sharing in its exploitation" (Christie and Scott 1965:6). However, the idea of "common property" is double-edged in that whales may be seen as belonging to nobody (*res nullius*) or to everybody (*res communis*). It is probably not unfair to say that those nations wishing to exploit whale resources have regarded them as belonging to nobody, while those which object to whaling see them as the "common heritage of mankind" (Hoel 1986:5-6), and that these are the two lines of battle now drawn up at the annual meetings of the IWC.

The IWC is open to any nation, regardless of its ability to exploit whale resources, and owes its origins to the International Convention for the Regulation of Whaling (ICRW), organized by the United States in 1946, although there were earlier statements on the need for international cooperation in the management of whale resources.[12] Its

---

[12] For example, the International Council for the Exploration of the Seas (1925), the Norwegian Whaling Act (1929) and the establishment of the Bureau of International Whaling Statistics, together with the International Whaling Convention (1931).

aims—like those of many national laws for fishery management today—can end up by being basically contradictory since they try to maximize both biological and economic returns. On the one hand, we find that the IWC exists to "protect all (whale) species from overfishing", and, on the other, that it is designed to arrange the "orderly development of the whaling industry". But it may well be that what has been wrong is not the goals as such but the ways in which they have been implemented (Andresen 1989:102) for, during the early years of the Commission's existence, members were more concerned with the development of the whaling industry than with the biological need to protect large whale species in the Antarctic. Thus between 1949 and the late 1960s, the IWC could be categorized as little more than a "whalers club", at which those nations heavily involved in whaling fixed an annual quota of whales to be caught. This quota—like the original motivation for the holding of the ICRW—was based not on ecological concerns, but on limiting the production of whale oil, so that for the next two and a half decades the IWC used the Blue Whale Unit (BWU), based on the amount of oil produced by a blue whale, as a standard measure for regulating whaling as a whole (Hoel 1986:49).[13]

Clearly, such a method of resource "management" encouraged whaling nations to overfish and bring some stocks near to extinction. The fact that the total quota system encouraged keen competition, with everyone trying to flense for themselves as many slices of the "whale cake" as possible, led to what is known as the "Whaling Olympics". The system encouraged, too, a reckless investment in larger and more efficient fleets (especially catcher boats) and to the overcapitalization of the pelagic whaling industry as a whole. Worse, it also meant that more whales were brought to the flensing decks in less time than ever before and. As a result, they were not processed with due care so that there was great wastage.

The BWU was abandoned finally in 1971, when the IWC turned to managing whales species by species in the Antarctic, before implementing a new management procedure of examining whales stock by stock in 1975. This classification was to be carried out by a Scientific Committee (Hoel 1986:107-108). These steps might have

---

[13] Incidentally, catches in the Antarctic were not reduced in accordance with scientific advice until the whaling nations were unable to catch their quotas (Andresen 1989:105).

provided an adequate system by which whale stocks could be sensibly monitored, but by this time whaling had become a moral issue which affected the membership of the IWC itself. From being fundamentally a "whalers' club", the IWC became during the 1970s an organization which was dominated by non-whaling nations, so that—by 1982—28 of its 39 members did not whale.[14] Among these new members we find such countries as Antigua, Belize, Costa Rica, Dominica, Mauritius, Monaco, Oman, Santa Lucia, and the Seychelles. It would seem, moreover, that some of these nations have joined the IWC solely in order to help bring about the moratorium. For example, anti-whaling activists, mostly based in Miami, first drafted required membership documents for at least six of these nations, then paid their annual membership fees (plus all-expenses-paid trips), and finally supplied them with suitably "green" commissioners (Spencer et al. 1991:177). By participating in the IWC debates, such small countries have managed to get an international visibility that would otherwise have been impossible (e.g. Ferrari 1983; cf. Hoel 1986:31, 76).[15] In the meantime, large countries like the United States and the United Kingdom make use of IWC meetings to appease their animal rights groups back home by speaking out against the "inhumanity" and "cruelty" of an activity in which they are no longer involved. The IWC meetings thus provide an ideal opportunity for such nations to kill (if we dare use the metaphor) two birds with one stone: politically, they can build up an international image as "conservationist"; economically, they have nothing to lose by protecting whales since they are not involved in whaling (Andresen 1989:109; Freeman 1990).

In the midst of the confusion at present characterizing the IWC, the Japanese have come to ask themselves a number of questions. For example, one point frequently made to us by both officials and whalers alike was that they could not understand why the hunting of

---

[14] Freeman (1990) has made a tri-partite classification of the development of the IWC: (1) 1949 to late 1960s: commercial decision-making; (2) late 1960s to mid 1970s: "scientific rationality"; (3) mid 1970s to present day: political decision-making.

[15] Their motives may not always be entirely obvious. Sperm whale oil has similar molecular qualities to a major export, copra oil, by which the Seychelles, for example, are anxious to protect their prime source of foreign revenue (Hoel 1986:78).

small type minke whales in Japanese waters (STCW) was prohibited, when the Alaskan Eskimos were allowed to harvest bowhead whales every year, even though at the time this species was considered to be extremely endangered.[16] The conclusion drawn from contradictions such as these is that there is a considerable amount of racial prejudice against the Japanese, who also suffer from the cultural imperialism of such countries as the United States and Great Britain. Not only this, but such prejudice is further linked to two other factors—the part played by Japan during the Second World War, and the present imbalance of trade between Japan and most western nations.

It is well to pause here for a moment. Are we to accept the idea prevalent among the Japanese that they are the victims of a conspiracy by western powers? Clearly, we cannot take this too seriously. After all, Iceland and Norway are publicly criticized for whaling, as are Canada and Norway for sealing, and Canada for trapping. But the Japanese will again ask which nations are particularly hostile towards them in the forum of the IWC, and here they seem to be on firmer ground. Anyone who had the dubious pleasure of reading what the British press had to say about the death of the Emperor Shōwa in January 1989 knows the kind of prejudice that is still harboured towards the Japanese with regard to their behaviour towards Allied troops in Southeast Asia during World War II. Indeed, a few years ago a Japanese naval visit to the Netherlands had to be cancelled because of strong anti-Japanese feelings still prevalent there. Some people wonder whether it is only coincidence that it is the United Kingdom and the Netherlands which have consistently attacked the Japanese for their whaling activities.[17]

Another culprit, so far as the Japanese are concerned, is America—the nation which has most to lose through Japanese domination of international markets, and which appears to have made it its business also to keep up the offensive on Japan's delegation at the IWC meetings in recent years. Why have American environmentalists, they ask, made such a fuss over the killing of a few hundred

---

[16] Neither the FAO nor the IWC's Scientific Committee supported the decision for a world-wide moratorium on commercial whaling made in 1982. The latter has, however, opposed aboriginal whaling of bowhead whales.

[17] One more footnote to this chapter of ironies is that the major opponent to the reduction of quoats between 1952 and 1962 was the Netherlands (Andresen 1989:104).

dolphins on the Japanese island of Iki, for example, when every year tens of thousands of dolphins are taken by American tuna seiners before being thrown overboard without being eaten? How come the American attitude towards whaling hardened considerably once the US delegation was able to put through the IWC its special request for limited bowhead whale hunting by the Eskimos? Here a number of Japanese feel that their support then was betrayed when the expected trade-off in quota numbers for other species of whale was not fulfilled. Again, why should they not be allowed to carry out research whaling, when such a kind of whaling is even provided for in the IWC's articles? Yet, the United States, which provided a precedent by its own research whaling in the 1960s (Holt 1985:198; Hoel 1986:82), now claims—along with other "environmentally concerned" countries—that the Japanese are not doing "research" at all, but conventional whaling in disguise.[18] Further coincidence to add fuel to the growing Japanese suspicion of a "conspiracy theory" against them is the fact that the United States, Great Britain and the Netherlands were, until recently, all active whaling nations. Japan's prominence as a whaling nation has been built on their decline.

Perhaps these underlying grudges might have surfaced in more orderly manner if, on the one hand, members of the IWC had not witnessed the seemingly outrageous behaviour of Japan's loggers and fishermen in various parts of the world;[19] and, on the other, the IWC delegates had not been so quick to attack Japanese views. As one commissioner from a European country said to one of us: "Many

---

[18] Gulland (1988:45) suggests that the Japanese are cloaking commercial whaling in the concept of "scientific whaling", but at the same time recognizes that surveys of stocks and other kinds of scientific research can only be conducted in association with commercial operations because of the costs involved. From the Japanese point of view, however, their scientific whaling is far from commercial. Not only is the whaling itself subsidized by government and private donations; the meat obtained is sold at government, rather than market, dictated prices (Nagasaki 1989:37).

[19] It is hard to say precisely whether the Japanese are in fact worse than others in this respect. Being the only non-whites involved until quite recently, they have made ideal scapegoats for almost every newly recognized environmental "disaster". At the same time, however, we must not forget that the Japanese are comparative latecomers to the exploitation of the world's natural resources and hence tend to have found themselves responsible for their final exhaustion. We should probably ask ourselves, therefore, about the role played by those other (western) industrial nations that started these processes of exploitation.

western delegates oppose or throw suspicions on whatever the Japanese say at the IWC meetings". To some extent victims of excesses by their fellow countrymen, and certainly victims of Japan's economic success internationally, those whalers who are left now find themselves totally helpless and for the most part without friends.

This may cause problems. As we are at pains to point out in this book, there is in Japan what can be called a whaling culture—a culture which is now fast becoming a symbol of national pride. Take food, for example. It is being argued that the Japanese have been eating whale meat for over 1,000 years, whereas they have been eating beef only since the late nineteenth century. Yet, the British tabloid, the *Daily Star*, on May 11, 1991, announces on its front page: *Sickest dinner ever served: Japs feast on whale*, and over two inside pages, complete with suitably gory pictures, readers are told about how *Greedy Japs gorge on a mountain of whale meat at sick feast* in a *banquet of blood*. Not surprisingly, the Japanese react with dismay to such reports—and the anti-Japanese feelings behind them—which they regard as an insult to their national culture. Hence, reactions such as the following two comments are hardly surprising. "The Americans demand that we stop eating whale meat and, instead, consume American beef. That is an insensitive and selfish demand."[20] or, more forcefully, "here we have the opinion of one race forcing its ideas on the traditional eating habits of another."[21] Given the various pressures exerted elsewhere on the Japanese (opening up their domestic markets, increasing the proportion of their GNP on "defence"), there is reason to believe that a new nationalism might emerge in Japan. Already, we find Japanese chiding their government for be.ng "timid and lazy", and for not "dealing seriously with the opposition in the IWC". Such comments as "Japan is the only nation in the world still consuming whale meat and their savage act is not excusable" by a US senator are thus widely publicized and condemned.[22]

Our major concern in this book is with the more specific culture of Japanese whaling and, in the following chapters, we argue that vital skills are being lost, together with the various values that accompany

---

[20] *Yomiuri Shimbun*, November 10, 1982.

[21] *Nihon Keizai Shimbun*, September 2nd, 1984.

[22] *Whaling and the Other Side of the Coin*, p.23 (Japan Whaling Association).

those skills, as a result of the moratorium on whaling. We are thus concerned, not with strictly economic or commercial problems, but with the question of culture as a whole. In treating this theme, we will describe some whaling communities in Japan, before giving a portrait of those who live there and describing the history of Japanese whaling as a whole. From this we will move on to the way in which work is organized, making important distinctions between the various types (large type coastal, pelagic and small type coastal whaling) and sets of activities (hunting, processing) in Japanese whaling. This allows us to argue for an "integrated whaling culture" and, in the next chapters, we show how such a culture is maintained through recruitment and career patterns, on the one hand, and through local community customs, on the other, with whaling companies acting as the vital link between the different types and sets of activities found in whaling in Japan. Finally, we look at the way in which this integrated whaling culture has been threatened with extinction by the present moratorium.

CHAPTER 2

Japanese Whaling Communities

In this chapter, we will introduce some of the Japanese communities which have, until recently, been heavily involved in whaling. Our aim here is to give our readers a rough idea of what it is that makes both whalers and their communities so special; in other words, of what makes them different from other types of community in Japan. Perhaps our first task, then, should be to define what we mean by the concept of "whaling community" in the context of Japan.

This is by no means as easy a task as it may at first glance seem. Initially, it would seem reasonable to define a "whaling community" as an administrative unit (such as a village, town or city) which has a land station where whales are brought to be flensed and processed into various products. Such a definition would, however, immediately leave out those communities without landing facilities, but which have, nevertheless, depended heavily on whaling—particularly pelagic whaling—as a source of employment and income for their inhabitants. Given this problem, and accepting the *Encyclopedia Britannica's* definition of a community as "a group of people with common characteristics and living together within a larger society", we can probably best characterise a whaling community as a group of people directly or indirectly involved in whale related activities (such as the catching, flensing, processing, and/or marketing of whales and whale products), and for whom whale related activities are important elements in the establishment of their self-identity. Very often, but not always, a whaling community is centered on one or more whaling stations in the vicinity. Defined in this way, it can embrace the greater part of a township (in which case, of course, we can talk about a "whaling town"), or only a small fraction thereof. A community is thus defined by its economic, rather than purely residential or administrative, criteria, and in this respect a whaling community is close to the classic concept of a "farming", "potting", "fishing", or other type of community usually found in Japanese society (Nakane 1967; Kalland 1981; Moeran 1984).

As we will show in Chapter 4, whaling has had a long history in Japan, and whaling stations were once widespread throughout the country (see Map 1 and 2). What these maps fail to show, however, is the gradual shift that occurred in the concentration of stations from the southwest to the northeast, where many new stations were established during the first decades of this century. As most of the whaling activities moved northwards, whaling ceased to be of importance to some of the old whaling communities in the southwest of Japan. Some of the latter—for example, Taiji, Arikawa and Ukushima—were nevertheless able to adjust to the new situation and their members managed to continue to be involved in whaling. It is these centres, together with the newer whaling communities of Wadaura, Ayukawa and Abashiri further to the northeast, that we will introduce briefly here.

Abashiri

Abashiri—a five hour's train ride from Hokkaidō's capital city of Sapporo and also connected by daily direct flights to Tōkyō—is the northernmost of the whaling communities found in Japan today. It is proud of its location in the far north, and all visitors are welcomed to the "City of 44° North", which—in comparative light—means that Abashiri is located on the same latitude as Nice in France. Even so, a visit to Abashiri in the middle of winter can be a chilling experience, and Abashiri's tourist industry cleverly exploits the near-arctic conditions. The main attraction, drawing about one million tourists during mid-winter, is the drift-ice which blocks the harbour—and forces boat-owners to beach their vessels—for about two months every winter. This is also the time when the snow festival takes place. Those visitors who miss the drift ice and arrive in Abashiri during the summer months—or during an exceptional winter without ice—may visit the "Drift-ice Museum", beautifully located on a hilltop overlooking the city and the Okhotsk Sea.

Despite the generally adverse weather conditions, Abashiri City has grown into one of the most important fishing centres in Hokkaidō, and can today boast a stable population of about 43,000 inhabitants.

Of these, 508 persons were directly engaged in fisheries in 1985, including eighteen full-time whalers (Iwasaki 1987:24). Although several attempts at whaling were made in Hokkaidō during the nineteenth century, Abashiri first became involved in whaling in 1915 when Tōyō Hogei (the leading whaling company in Japan at the time) opened a land station for large type whales. Three years later it was closed because of its inconvenient location, but in 1931 a new land station was finally established in a better position. This station remained in operation with one interruption—no whales were brought to Abashiri from 1936 to 1939—until World War II. Another company, Tosa Hogei, then opened a second station in Abashiri and 30 whales were processed there in 1940. After the war, however, only Nihon Suisan (which had absorbed Tōyō Hogei) retained its LTCW station in Abashiri, and this, too, was finally closed in 1959.

The outbreak of the Pacific War in 1941, meant trouble for the Japanese whaling industry, since most of its whaling fleet was taken over by the authorities to be used in the war effort. At the same time, however, whales increased in importance as a potential source of food, so that coastal whaling was then given special consideration. Quite a few new small catcher boats were brought into operation. Most of these were owned by the large whaling companies (or by their subsidiaries) which had been engaged in pelagic whaling before the war. For example, Nihon Suisan (or Nissui as it is commonly known) bought five boats, which were transferred after the war to subsidiary companies—one of them being Hōyō Hogei which operated out of Abashiri.

By the end of the 1940s then there were at least six minke whaling boats based in Abashiri. Three of these vessels were owned by Hōyō Hogei, another two by Hokkaidō Hogei (a subsidiary of Nihon Reizo), while the last was owned by Hokuyō Hogei. Ownership changed radically during the following decades, however, as the number of boats declined due to licensing policies, and those that remained were in the end taken over by individual owners. Today there are only two STCW boats left in Abashiri, both of them licensed to catch small type whales—in other words, minke and Baird's beaked whale. One is owned by Miyoshi Hogei, which was established when the founder acquired one of Hokkaidō Hogei's boats; the other by Shimomichi Suisan, which was set up when its founder bought a boat from the company that had replaced Hōyō Hogei, Aoshima Suisan.

Most of the whalers working in LTCW came from southern Japan,

Map 1: Location of some Japanese Whaling Communities

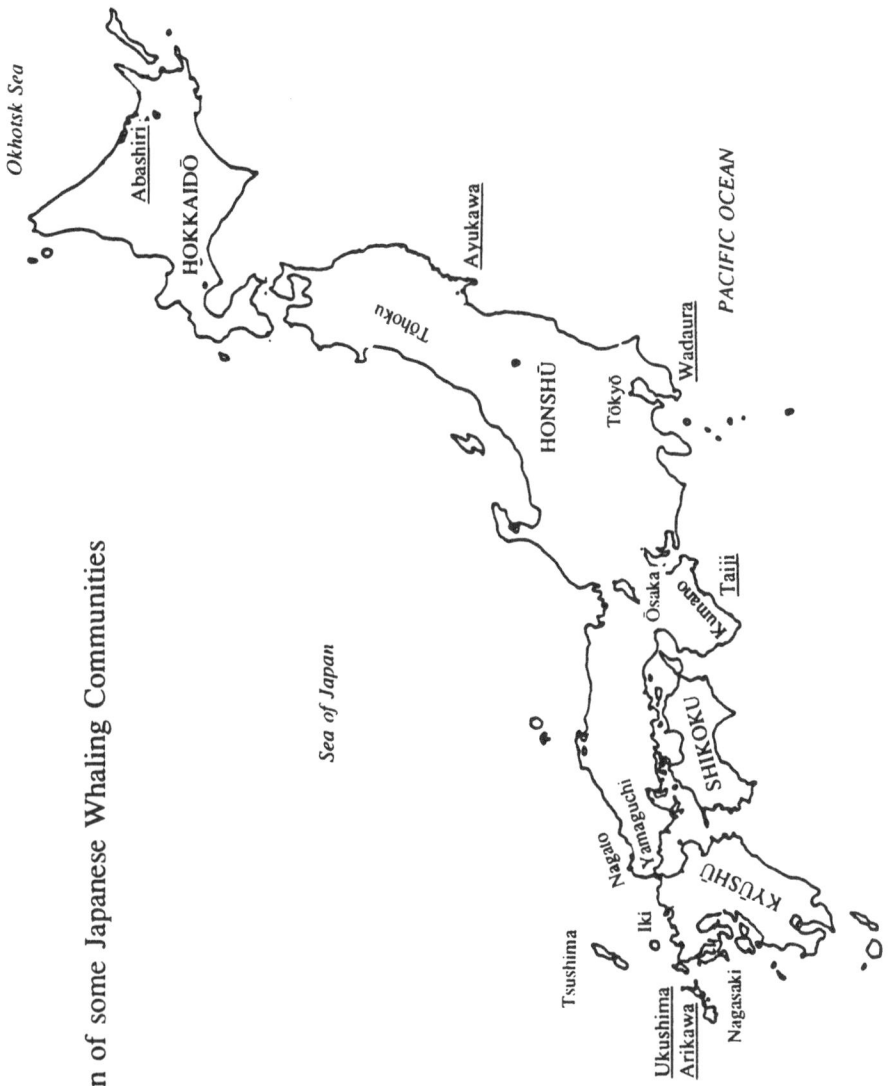

since the whaling companies used to bring boat crews, flensers and office staff to the station for the summer whaling season. During the rest of the year only a bare minimum number of employees was kept on in Abashiri to take care of the facilities, but those companies which bought whale residuals (bones, entrails and blood used mainly for the production of fertilizer), and which were based in Abashiri, did employ local people. The crews on the STCW boats tended not to be local either, but to come from the main island of Honshū or further south, and we find that the founders of the two existing STCW companies in Abashiri originally came from Fukui and Miyagi prefectures, while the present gunners on the two boats hail from Miyagi and Kagawa prefectures. Nevertheless, the proportion of locally born whalers has increased and at present all but a handful of the whalers live in Abashiri with their families. During the summer of 1987, which was the last normal whaling season before the moratorium went into effect, altogether sixteen persons worked on these two boats, while another sixteen worked at the two flensing stations on land, either as office workers or as flensers (Iwasaki 1987:37).

Although other people in Abashiri have been involved in marketing the whale meat, whaling has in recent years been of limited importance to the town as a whole. Indeed, even in its heyday, it is unlikely that whaling in any way dominated the town's economy, so that it is difficult to call Abashiri a "whaling town" as such. Nevertheless, within this larger residential area, there is a group of people for whom whaling-related activities have been of fundamental importance—both as a way of livelihood and to their sense of self-identity. In this respect, there can be said to exist a whaling community. Iwasaki Masami, who has studied the whalers there over several years, states that this whaling community consists of a core whaling group (i.e. the owners and those who hunt the whales), together with factory workers who are involved in processing the whale products, office workers, wholesalers, and retailers, plus their dependents. "The role of each of the members is defined, and the whaling community remains cohesive and discretely defined in the larger local community of Abashiri" (Iwasaki 1987:39).

The two small flensing stations in Abashiri, placed side by side and sharing the same slipway, have been closed. They looked alike, had been built by the local administrative authorities, and were rented by the two companies. Until the stations were closed flensing of

Baird's beaked whales was done outdoors on wooden floors, so that at each station a small shed for housing work tools was the only building. Minke whales, for their part, were roughly flensed at sea, and the meat was trucked directly from the boats to worksheds located further inland.

Ayukawa

Ayukawa is in fact one of a number of hamlets making up the township of Oshika, situated at the end of a long peninsula stretching down into the Pacific Ocean from Sendai, the capital of Miyagi Prefecture. With a population of about 2,000 inhabitants, Ayukawa has expanded considerably from the fishing village of some 80 households, with fewer than 500 people, that it was before the advent of whaling in 1906. Three years later, there were 110 households, inhabited by 651 men, women and children, and thereafter there was a steady influx of people from all over Japan. Statistics show that in 1946 there were 2,900 people living in 540 households, while in 1955 3,795 people lived in 729 households.

The real meaning of this decline in population becomes more obvious when one traces over the years the increases and decreases in age groups in Oshika Township, of which Ayukawa is a part. Of the 1,885 children aged between 5 and 9 years in 1955, for example, only 537 remained in the town when they were in their late 30s, while during the same period only 596 of the 1,325 people who had lived there in their early 20s were still there thirty years later. At the same time, although the number of young people has drastically decreased, the overall percentage of old people has been on the increase. It is those who are over 50 years old who form the bulk of the township's present population. Not surprisingly, many of those now living there see Ayukawa as a dying town, for they remember all too well the heyday of whaling when almost 4,000 people lived there. In those days, they fondly recall, the bay and the narrow straits separating Ayukawa from the island of Ajishima were packed with ships, and the streets were filled with people. Then there were inns, bars, the occasional brothel, and all kinds of entertainment in a town which thrived on a thriving industry.

Now things are different. The houses are still there but for the most part they look drab and deserted. Many of them clearly cost a lot of money, built in the days when whaling meant prosperity, and the visitor might be forgiven for thinking that the community is quite well-off. There is one large hotel facing onto the street that was once the waterfront, and busloads of tourists come to stay here at weekends, on their way to or from the sacred (and sparsely populated) island of Kinkazan—half an hour away on one of the ferries that ply back and forth between Ayukawa and this famous tourist site. Along the same street are lined a number of large shops and restaurants which sell shellfish, seaweed, abalone, all kinds of fish and whale crafts to those tourists who make the long drive down the Oshika pensinsula from Ishinomaki, more than 40 kilometres away.

In spite of its remote location, Ayukawa, it seems, has only tourism with which to keep alive the hopes and purse-strings of its remaining inhabitants. There is a gateway with whales on it spanning the road that winds down the hill from the primary school at the end of the bay. Further down into this *kujira no machi*, or "whale town", the fire station sports four cute whales on its façade and beside the now closed whale museum on the waterfront, there are two iron whales which are used in the "whale festival" held every August. Beyond the ferry pier is a whaling vessel, which has been dragged out of the water, repainted sky blue and maroon red, and donated to the village by one of the whaling companies that used to operate in Ayukawa. It is the main attraction of "Whale Land", which was opened in October 1990. At least, those who come here know what Ayukawa stands for, and, if they look across the harbour with its maze of masts of fishing boats, they can still see the flensing stations of two of those same companies. Physically, Ayukawa is very much a "whaling town".

With the moratorium things are changing. There are only two flensing stations still operating, and one of these is so small and insignificant that one could easily miss it. Nowadays, three STCW vessels are registered in Ayukawa and there are still times during the summer months when the chug-chug-chug of a boat late at night will draw a few dozen spectators who come to watch the flensing of a Baird's beaked whale. At the same time, however, the fact that it is only occasionally that such activity takes place merely serves to remind whaling familes of the "good old days", when whales were plentiful and people had money in their pockets and plenty of whale for food on their tables.

One concomitant of the moratorium is the unemployment that

affects those living in Ayukawa—particularly, as we earlier implied, young people. This means that the town is now populated for the most part by people who are well into middle age. Sure, there are fishermen with their families, and quite a few who live off the tourists, but any child going on to high school finds his or her self having to commute almost as far as Ishinomaki every day, or else leaving home to live in lodgings nearer school. Not surprisingly, with money so short, a number of parents are moving out with their children. Soon, it seems, there will be very few left. Ayukawa will become a small fishing community again.

Wishful thinking? The Oshika Town office, which is located at the back of Ayukawa, has other plans. Having already arranged for a scenic driveway, known as the "Cobalt Line", to be put in along the top of the hills that form the spine of the peninsula from Ishinomaki to Ayukawa, the local government has been involved in creating what it has called in English, the "Whale Land". In the process of negotiating enormous loans in order to complete this project, those concerned were desperately trying to rescue their town from the effects of the moratorium. In a few more years, it seems, Ayukawa will be nothing more than a tourist resort which, like most other places with a past, will be obliged to invent its "traditions".

And in what precisely does Ayukawa's past consist? As with many other villages established in the middle ages, Ayukawa folklore suggests that the earliest families to settle there were either connected with noble lords on the run during the period of civil strife between Northern and Southern Dynasties (1336-1392), or with ten samurai who took refuge there during the Civil Wars of the 16th century. Whether either tale is true is more or less impossible to tell, but it seems that the dozen or so families who arrived way back in the distant past divided up the sea front between them, starting with Ōmori in the north and finishing with Endō in the south, and that together they worked a village fishing net. As time went by, these families created branch households, offering younger children a place to stay in exchange for employment on the land and working the nets. Outsiders, too, were brought into community life this way, so that eventually there emerged a hierarchical structure in which the oldest families (known as *amimoto*, or net owners) employed the other families as *amiko* (net workers).

All this changed to some extent with the arrival of the first whaling company, Tōyō Hogei, in Ayukawa in 1906. In fact, it was the "big men" of the village—the *amimoto*—who, realizing that fishing was on the decline, persuaded the other villagers to allow Tōyō Hogei to establish a land station in Ayukawa. In the course of the next few years, local inhabitants suddenly found themselves comparatively well-off, but none more so than the old *amimoto* who cleverly shifted from net fishing to the production of fertilizer, gaining a monopoly on the supply of whale bones, entrails and blubber, and employing their former *amiko* as wage labourers in their factories.

Not that the old hierarchical structure of the village came to an end there. It was true that the net workers now had plenty of cash, and some choice of what kind of work they could do, but the old families remained in control of Ayukawa's affairs and more or less monopolized the village headship, as well as that of the fishing cooperative. It was these families—people like Izumi Tsunetarō and Ōmori Takijirō—which then decided in 1925 to establish the village's own whaling company, Ayukawa Hogei, when they were threatened with loss of supplies for their fertilizer production plants. Thirty shares were issued and bought up by the new capitalists, who then proceeded to order a new boat, but their entrepreneurial plans were upset by the fact that prices for whale meat were falling rapidly, demand for fertilizer was also decreasing, and initially the company was only allowed to fish for sperm whales far out at sea. In such circumstances, it was hardly surprising to find the company being bought up by what then became Kyokuyō Hogei in the mid-1930s, although the Ayukawa whaling company with its predominantly local workforce remained more or less a separate entity therein until the end of the Second World War.

Although the new wealth created by whaling eventually broke down the old "feudal"-like relations between *amimoto* net owners and *amiko* workers, the old families were able somehow to continue through the years as the Ayukawa elite. In recent times, we find that they have shifted to operating ferry boats to the neighbouring island of Kinkazan, running restaurants and souvenir stores for tourists, while still maintaining their interests in fishing. They, together with a few of those who were employed by the whaling industry, are the ones still to remain in Ayukawa now that whaling has come to a stop. Thousands of others have moved elsewhere, leaving what was once a thriving whaling community with no more than the bones of their past. It is these that fertilize their collective memory, these that fuel the hope that, one day, Ayukawa will be permitted to go back to STCW.

Where once people dreamed of "progress", seeing their future lit by the success of the whaling industry, they are now left with the nightmare of the past and with nostalgia for what might have been.

Wadaura

The town of Wada is located in Awa County, Chiba Prefecture, on the Pacific coast of the Bōsō Peninsula, just south of Tōkyō. The southern part of the peninsula is hilly, but with small patches of flatlands along the coast with its short rivers, and these are intensively cultivated. Flowers for the Tōkyō market have become a speciality of the region. The shore is rugged and there are few natural harbours. Rocky shallows extend far out to sea in places, making this area a good breeding ground for abalone, and women abalone divers—famous both at home and abroad—have operated along these shores for centuries. But the rocky seashore also has its drawbacks. There are few stretches of sandy shore where beach seines can be used in the way they are at Kujūkurihama further to the north. Moreover, sea voyages have always been a dangerous undertaking in this environment, and many sailors and fishermen have found a watery grave along this treacherous coast.

Wadaura, the administrative centre of the township, is the most accessible of all the whaling communities in Japan, for it can be reached in less than three hours from downtown Tōkyō. Despite this, the area has a rural appearance and there is little here of the kind of industrial development that mars the Kantō region. This has made this coast a kind of resort for Tokyoites, many of whom have bought flats in high-rise apartment buildings facing onto the sea. Nonetheless, Wadaura gives even the casual visitor the impression of being a fishing town. The houses are squeezed into a thin strip of land separating the sea from the steep hills beyond. Narrow streets wind their way within a patchwork of wooden houses.

A highway separates the cluster of residential houses from the harbour, the center of the village's economic activities. In one corner of the harbour, which is protected by large breakwaters, lies the flensing station of Gaibō Hogei. It is not a very impressive sight—little more than a concrete slipway, most likely covered by seaweed laid out to dry in the sun. There is a wooden roof built over the flensing floor, and behind it are the remains of a winch that was once used by the LTCW company Nittō Hogei to drag sperm whales up the slipway. No measures were ever taken to keep on-lookers out until foreign anti-whalers hired a lodging (*minshuku*) on the other side of the highway and followed the whalers' every move through high powered binoculars. Naturally annoyed by what they perceived as an invasion of their privacy, whaling companies then set up screens beside the slipway in order to prevent further observation by outsiders.

Wadaura is the most recent of the whaling communities discussed in this chapter. It was in 1948 that the newly founded STCW company Gaibō Hogei was established and opened its station in Wadaura. All the same, the taking of Baird's beaked whales has an ancient history in this region, for whaling was first organized by an old and influential family in Katsuyama—situated across the peninsula by Tōkyō Bay—as early as in 1612. At that time, whales were surrounded by a large number of fishing boats and attacked with harpoons. In the early part of this century, however, modern Norwegian harpoon guns were introduced, and the old land station was moved from Katsuyama to nearby Tateyama, and later still to Shirahama and Chikura, before finally ending up at Wadaura (Akimichi et al. 1988:86). At present Gaibō Hogei's two STCW boats are registered in Wadaura and, together with a third boat, they have until 1988 landed 35 Baird's beaked whales there every summer. The LTCW company Nittō Hogei opened a second land station, using the same slipway as Gaibō Hogei, and caught sperm whales—which were brought to Wadaura in order to be flensed—during the winter months. In summer the company caught Bryde's whales, but since most of these whales were caught and processed at their station in Ogasawara, conflicts over the use of the slipway and flensing floor were minimal. Nittō Hogei's station was destroyed in 1987 when the moratorium was enforced.

Both Gaibō Hogei and Nittō Hogei have brought most of their whalers into Wadaura from elsewhere. Whalers from Ayukawa replaced local crews on the two STCW boats operated by Gaibō Hogei, once it started minke whaling in the 1970s. The master flenser, too, comes from Ayukawa, but the other flensers are locally employed. Both the crews and the flensers employed at Nittō Hogei have come from a wide area so that only a handful of the 3,008 inhabitants of Wadaura (1985) is actually directly involved in whaling. The impression given, therefore, might be that whaling has been of limited importance to the local economy and identity, but this would be wrong for at least two reasons. Firstly, although whaling is rather new to Wadaura itself, we have seen that whaling has been carried out in Awa County for a number of centuries, so that Wadaura can be regarded as a part of a larger whaling complex comprising the entire southern area of the Bōsō Peninsula. Much of the meat from the Baird's beaked whales is sent to Chikura and Shirahama nearby where it was marinated and sun-dried, and this product, *tare*, is one of the favourite foods of those living in the area as a whole (Akimichi et al. 1988:27). Fresh meat is also distributed throughout the southern part of the peninsula, including the old whaling town of Katsuyama.

Secondly, whaling has been of great financial importance to Wadaura. The land for the flensing station has been rented from the local Fishing Cooperative Association (FCA), and in 1984 Nittō Hogei paid ¥7.8 million in rent for five winter months, while Gaibō Hogei paid another ¥7.7 million. Both companies also bought large amounts of ice from the FCA (Nittō Hogei spending more than ¥9 million on this in the 1985/86 season), so that the amounts paid by these two whaling companies have gone a long way towards subsidizing the FCA and Wadaura's fishermen. Whaling also stimulated other activities, and the many whalers coming into Wadaura for the whaling season required accommodation, as well as restaurants and bars in which to eat and drink. Wadaura has been, in short, a whaling community.

Taiji

Taiji is a small town of about 4,000 inhabitants. It is set among picturesque hills on the rugged Kumano coast in Wakayama Prefecture, somewhat off the beaten track. Nowadays there are roads and railways passing through Taiji, but it takes the better part of a day to get there from Nagoya or Ōsaka for driving is a slow process along the coast road which twists and winds its way past numerous capes and inlets. Despite its remoteness, however, there have always been a lot of travellers—mostly pilgrims—in the area, which is famous for its places for ascetic practices. In the old days, they walked and had the time to chat with the locals, or watch their whaling activities, but nowadays they are ushered around in tourist buses that tend to bypass people. Several new species of travellers have emerged in recent years—the dedicated sports fisherman, for example, who hauls himself off to the remotest point or else hires a fishing boat; or the city businessman who spends a short weekend at an inn, gets drunk, makes passes at the establishment's female employees and for a moment believes that he is back in the "real Japan". For Taiji is a typical *furusato*, "the good old village back home".

Still all *furusato* need an image of one sort or another. Further up the Kumano Coast, the female divers (*ama*) provide that image, and busloads of tourists are brought to the Mikimoto Pearl Island to see female divers dressed in white see-through blouses take up pearl oysters. That the real *ama* are dressed in black wet-suits and dive for abalone does not seem to matter (Martinez 1990). In Taiji, it is whaling that has given the community its identity, and the town has done a lot to foster that image (Takahashi 1987). A whale museum was opened in 1969, its façades covered with mosaics of whales. For an extra 1,000 yen, visitors can also board a beached catcher boat which was once used in the Antarctic. Statues of dolphins adorn the bridges, a gateway with a whale has been built and whales are cast on all the manholes in the township, while one even tops a local *pachinko* pinball parlour. A hotel, jointly built by Taiji Town and the neighbouring Katsuura Town, uses a cute dolphin as its emblem. In fact, the hotel's towels, with their whale emblem, became so popular with visitors who stole them that the hotel can no longer afford to provide towels for its guests, but sells them instead in its main lobby. The town has also built, in concrete, a look-out post at the place where whales used to be spotted in the old days. Although visitors' chances of sighting a whale are slim, the view is spectacular. Not far from this view-point is a large sculpture of a right whale. Here, memorial services (*kuyō*) for whales are held annually by the Taiji Whalers' Association, in cooperation with the town, on 29 April. Elements of this festival are, like the festival in Ayukawa, newly created in order to attract tourists, and it is thus one example among hundreds of local festivals which are invented to provide a special identity for a *furusato* (Bestor 1985).

There is then a certain potential for tourism in Taiji, and this industry provides some local people with a reasonable income, but in order to survive in this competitive market, the town has to live up to its image. This means that if Taiji is to continue to be seen as a whaling town, it must be able to let the visitors meet and talk with active whalers as well as eat whale meat at their hotels, rather than simply gaze at invented festivals and whale-images on manholes. Even a long and glorious history of whaling, nicely explained through museum exhibits, can hardly make up for the lack of active whaling.

Tourism is a risky business—especially when the very corner-stone on which it is based is cracking. Worse, there is hardly any farm land in the vicinity and Taiji's distance from central markets makes it difficult to attract industries, except the polluting kind that nobody else in Japan wants. This means that most of the people in Taiji have little else to do but to rely on the sea, as they have always done.

Taiji is the traditional whaling town *par excellence*. This is the "town which lives with the whales" (*kujira to tomo ni ikiru machi*). It claims an 800 years' history in active whaling, and it is here that the first organized large-scale whaling in Japan is believed to have been initiated when, around 1606, Wada Kakuemon—Taiji's great "son" and reputedly descendent of an ancient warrior family—organized a large whaling group to hunt such baleen whales as humpbacks and right whales with harpoons. This method of whaling spread rapidly throughout southwestern Japan, only to be replaced in 1675 by the more efficient method of net-whaling that was also invented in Taiji by one of Kakuemon's descendants. (A detailed description of this mode of whaling is given in Chapter 4. See also Taiji 1982:3-29.)

Whaling brought prosperity to the people of Taiji during the Tokugawa period (1600-1868), when they hunted large type whales with nets during the winter months and pilot whales with harpoons during summer. But as the Tokugawa period came to an end, catches of right and humpback whales declined, probably as a result of the whaling activities of the Americans and other westerners in the Pacific. Consequently, Taiji was plunged into recession—a recession so bad that, in 1878, poverty drove whalers into breaking a long observed taboo of never hunting a female right whale with calf. The result was that at least 111 men lost their lives in a fierce storm (see Frame 8).

This disaster caused the final collapse of net-whaling in Taiji, and attempts to revive the large whaling enterprise failed. Although the importance of pilot whaling increased, and a new five-barreled harpoon-gun for this kind of whaling was developed in 1904, many young men left the area for Hawaii and the American continent. Only the introduction of modern methods of catching large whales in 1905 brought new prosperity to Taiji. In that year, Tōyō Hogei started whaling from Taiji with a catcher boat captured from the Russians. In the same year, too, a local company—Kumano Gyogyō—started whaling, and three Norwegians were hired as gunners and instructors. They impressed the local people by the plug that they chewed almost incessantly and by the way that they spat a reddish saliva (Kumano Taiji-ura Hogeishi Hensan I'inkai 1965:163-164).

Not surprisingly, perhaps, Taiji whalers quickly became involved in modern whaling, in which they came to play a leading role. They travelled all over Japan in order to work at the newly established land stations in places like Ayukawa, and when the Japanese whaling fleets started going to the Antarctic from the mid-1930s onwards, Taiji whalers went with them as a matter of course, crewing on catcher boats, flensing on the factory ship, and working at anything else to do with whaling.

At the same time, Taiji men have continued to catch smaller cetaceans—dolphins and pilot whales. Shortly after World War II, there were twelve minke boats and eight boats catching pilot whales. Until recently, then, Taiji men have been involved in all types of whaling and in the heyday of whaling (around 1960), about 600 Taiji men were engaged in such activities. Since then whaling has contracted and when the land station in Taiji was closed in 1987, an epoch that had lasted almost 400 years came to an end. However, a few men are still engaged on the two STCW boats registered in the town, and Taiji men also take part in the research whaling at present being conducted (under international protest) in the Antarctic.

Whaling has made a lasting impression on Taiji in numerous ways. More than any other Japanese town, Taiji relied on revenues from this industry. In 1966 and 1967 whalers contributed two third of the town's total tax revenues and the town itself was without any debts. Compare this with 1984, however, and one can gauge the effect of the reduced quotas. In that year, only 4 per cent of the tax revenues were paid by whalers, while the town had more than one billion yen in debts (Table 23 in Government of Japan 1989). The town has again been plunged into a recession.

Nevertheless, there is a strong sense of community pride amongst the inhabitants of Taiji. This was first expressed after the 1878 disaster when the "Association for the Beloved Village" (*Aison dōshikai*) was established. More recently, the work of local historians has added further fuel to their self-identity, for these latter have documented how Taiji was a pioneer in the development of whaling methods in Japan, and argued that these methods were not imported from China as was previously thought (Takahashi 1987:163). After the Pacific War, three issues have strengthened this identity further: the attempts to merge Taiji with other townships, plans to construct a nuclear power-plant in the vicinity; and the international anti-whaling movement (Takahashi 1987:163-165).

In the 1950s, most Japanese communities were merged with others in order to create larger and more efficient units of local administration. Taiji successfully resisted any attempts by higher authorities to merge it with its neighbours, protesting partly because it was richer than its neighbours, but more importantly because of its fear of losing its identity as a whaling town. The second issue, that of the nuclear power plant, served to arouse Taiji people so much that plans for its construction have now been shelved.

The international movement against whaling has been more difficult for Taiji to combat, for it hits at the very foundation of Taiji's identity—whaling. At a time when the cultural heritage of whaling has been in the process of receiving wider recognition at home (Wada's grave had been made into an important historical monument, and in 1969 the revived "whaling dance" (*kujira-odori*) was designated an intangible cultural treasure by the prefecture), internationally foreigners tell them to stop whaling for they regard it as a barbaric custom. The net effect of all this has been the production of a whole new set of symbols in Taiji, which now includes the obligatory images of whales and dolphins that were mentioned earlier and which have emerged all over the township in the course of the last decade. The whale dance, too, has been made compulsory in Taiji schools and is performed on annual sport days. A group performing on "whale drums" (*kujira taiko*) has been established, and since 1985 a new "Old Village Festival" (*furusato matsuri*), which acts out whaling themes, has been held in the town (Takahashi 1987:160). Of all the whaling communities, only in Taiji have people banded together to form a united group, the Taiji Whalers' Association. Perhaps of greatest significance, however, the whale memorials (*kuyō*), which previously were performed at the temple on behalf of the whalers, have been made into public events. The memorial services, sponsored by the Taiji Whalers' Association, are held at the newly constructed whale statue on April 29th, the birthday of the late Emperor which is now, ironically, celebrated by the Japanese in general as "Green Day". In the event, they serve as a strong rallying call for the protection, not of whales, but of whaling. Everybody agrees that this is their only way of life in the future.

Ukushima

Ukushima is a small island at the northernmost tip of the Gotō Archipelago, off the west coast of Kyūshū in the south of Japan. The only way that it can be reached is by a ferry that leaves Sasebo every day, travelling from north to south along the island chain on one day, and from south to north the next. The journey can thus take anything up to five hours.

Ukushima itself is more or less round and fairly flat, apart from a large hill that rises in its centre and from which one can gaze down on the two main villages situated one each side of the island. The ferry docks at Taira in the east—a hamlet which takes its name from a member of the Heike clan who was washed ashore near there after the battle of Dannoura in 1185, and whose once elegant stretch of sandy beach is now being converted into concrete for a new ferry quay. There is also an island administrative office building and the local town hall, built right on the filled-in sea front, where the narrow road that goes right round the island broadens out to allow the light vans and occasional car to pass along the two lanes separated by an important-looking, but probably unnecessary, white line.

Behind the road are clustered the houses of those several hundred people who live in Taira. Like many fishermen's houses, their tiled roofs are frequently marked with a whitewash-like substance designed both to protect the roofs against winter storms and to indicate social status. Many of those who live there are, of course, fishermen, for the island is well situated to harvest abalone, small squid, and other sea resources in the unpolluted seas surrounding it. But there are also many who were whalers and who went with the Japanese whaling fleets to the Antarctic. Not only this, but Ukushima people are proud of the fact that a comparatively large percentage of those who became whalers managed to attain the enviable and prestigious position of gunner aboard the catcher boats. One of these men, now in his 90s, was gunner on the very first Japanese whaling fleet in the Antarctic, and he recounts with singular clarity—both in his memory and in his eyes—how they travelled to Antwerp to picked up the whaling vessel bought from the Norwegians, and then travelled south via South

Africa.

That Ukushima is clearly a whaling community may not be immediately apparent to the casual visitor. After all, not many of those who come to the island ask about who lives in Taira, or discover the intricate relations among whaling families who live in the clustered network of single story houses facing out onto the bay. Nor is there any flensing station on the island, to signify that this is a whaling community. Nor is there any other sign of whaling—apart from a memorial stone to the dead whales, set up by more than fifty whalers some thirty years ago.

But Taira is very much a whaling community, and islanders will proudly relate the folk legend of a mother whale that was hunted while travelling with its calf down past Ukushima to a famous temple on one of the southern islands of the Gotō Archipelago, and the subsequent tragedy that befell those who tried to kill her (Frame 8). The fact that this legend probably belongs more strictly to another island does not really matter to the people of Taira. After all, whaling is something that all those living in the Gotō Archipelago have been doing for centuries, and people readily recount the times when they have sighted whales out to sea. In short, whales and whaling are part and parcel of island life. Without them, Ukushima would not be the same.

Arikawa

Arikawa is located two ferry stops, or two hours, south of Ukushima. It is one of five townships on Nakadōri Island, also located in the Gotō Archipelago, Nagasaki Prefecture. Here we can still find unspoilt nature and clear water. "It is a good place to live and for the kids to grow up", people say. "There is less stress here and we are not so worried about making a career in life. We don't push our kids through prestigious schools either."

There are no such schools in Arikawa, nor in the Gotō Archipelago as a whole for that matter. Nor is there much in the way of a "career" to make. The best a man can aspire to nowadays is a position in the town office, although in the old days this was a fall back position for those who could not make a career out of whaling. There is not much industry either—a few fish-curing plants, and some Japanese noodle makers. As with most remote places in Japan,

Gotō—like Ayukawa—gets the industry that nobody else wants. In the neighbouring town of Kami-Gotō, for example, huge oil-tanks dominate the landscape—put there by a government that wants to be well prepared for the next "oil-shock". Their erection caused quite a stir on the island. What if one of the tanks cracked? But payments of compensation money are attractive to hard-pressed rural communities like Arikawa, and the ¥209 million which the town received over a five year period was extremely welcome. The neighbouring hamlet of Kōnoura was less fortunate. The local fishing cooperative association (FCA) allowed a construction company to extract stones from a nearby mountain against payments of compensation, money which was then invested in yellowtail nets. After a spectacular first catch, the catches of yellowtail have decreased ever since—as has the mountain, leaving an ugly scar in the midst of an otherwise beautiful landscape, and local people have come to regret the deal. "The capitalists are always clever at exploiting the economically weak", says our man at the town office with an embarrased grin.

Arikawa can only be regarded as pretty remote—especially by Japanese standards—although not as remote, perhaps, as Ukushima. It was this remoteness, together with the rugged landscape, which made it possible for Christian communities to hide away in Arikawa and practice their religion unnoticed when Christianity was outlawed in Japan early in the seventeenth century. It caused quite a sensation when they were "discovered" two and a half centuries later, after Christianity had again been legalized, although the fact that their interpretation of the Gospel had in the meantime deviated considerably from that taught them by the Jesuit missionaries caused some eyebrows to be raised.

Even with the development of modern means of transportation, Arikawa's competitiveness has not been increased in any way. On the contrary, the township has lost the few advantages it accrued during the heyday of shipping. True, it has a small airstrip and there are daily services to Fukuoka and Nagasaki, but the planes only take eight passengers and with about 30 per cent of the flights being cancelled because of bad weather, most people still arrive in Arikawa by the same ferry from Sasebo that calls at Ukushima.

Arikawa Town, comprising a mountainous peninsula with few open plains, has approximately 8,500 inhabitants (1990) living in 21 different hamlets. Not all of them, however, are inhabited by whalers who, for the most part, are strongly concentrated in the hamlets of

Enohama (where 79.2 percent of the income in 1973 derived from whaling), Ogawara, Akao, Kami-Arikawa, Hamaguri, and Naname. With the exception of Enohama, all of these hamlets are located on Arikawa Bay, where whaling conditions were once good, since the *Kuroshio* ocean current passes near by the coast, forming whirlpools in the bay and so favouring those species of fish like sardine and horse mackerel which the whales like to follow. This advantageous location has made Arikawa (as we will term all those hamlets that have been heavily involved in whaling) one of the oldest whaling centers in Japan.

Indeed, it would seem that whaling groups might have been established there as early as 1598, when Eguchi Jinzaemon is said to have started a group together with an entrepreneur from Kumano (where Taiji is located) who introduced the technology of hunting by harpoon (Nakayama 1987:147). At one time there may have been as many as ten whaling groups in Arikawa, with another eight groups in Uonome, a village located on the other side of Arikawa Bay. Many of the whalers came from other provinces, in particular Nagato (present Yamaguchi Prefecture) and Kii (present Wakayama Prefecture, where Taiji is located).

The new whaling technique of using nets—invented in Taiji in 1675—seems to have reached the Gotō Archipelago very rapidly, for a net group was organized in Uonome in 1678 and, three years later, Yamada Mobei from Ukushima was invited by the Gotō feudal authorities to introduce net whaling to Arikawa itself. From 1691 to 1693 Yamada and Eguchi Jinuemon (the fourth son of Jinzaemon) operated this net-group jointly, before Eguchi continued on his own (Nakayama 1987:147-148). The catches of whales—mostly right and humpback whales, but also fin and sei—reached a peak of 83 and 81 in 1698 and 1700 respectively, and by 1712 Eguchi's group had taken a total of 1,312 whales. After this the catch gradually declined, however, and between 1712 and 1727 only 496 whales were caught. Thereafter Arikawa whalers experienced economic difficulties and managed to continue only after a series of re-organizations (see Frame 7). The catches were particularly poor between 1789 and 1818 (Nakayama 1987), and the feudal authorities had to intervene and run the whaling operations during at least two periods. This instability of the whaling groups in Arikawa was shared by a number of other whaling communities in Kyūshū (Kalland 1986, 1988).

With the emergence of modern large type coastal whaling,

Arikawa lost much of its importance as a center for whale processing. Catches made off Arikawa had long been erratic, and the whaling grounds gradually shifted to Korea and the northern regions of Japan. New whaling grounds were also opened off Taiwan and, from 1922, Ogasawara (Bonin Islands). As a result, the land station in Arikawa was closed down in 1911, although it was re-opened for a few years in the early 1930s.

Nevertheless, this shift did not mean that whaling lost its importance for the inhabitants of the community, for a great many local people found employment in the new whaling companies and were sent to whaling stations all over Japan and its pre-war dominions. Arikawa men played a leading role in the modernization of Japanese whaling, and made several attempts to introduce new whaling methods from abroad. One of the most successful companies which used the American whaling method was Gotō Hogei (see Chapter 4), operating out of Arikawa, while the very first attempt at using the Norwegian method in Japan was made from Tainoura, in present-day Arikawa. Hara Shinichi, whose statue stands in the centre of the town, was a prominent figure in attempts to modernize the Japanese whaling industry, and he brought Arikawa whalers into the new companies. Like whalers in Taiji and Ukushima, the Arikawa whalers were engaged in whaling throughout Japan, taking with them knowledge of the new forms of technology to all corners of the country. And again, just like the whalers of Taiji and Ukushima, many Arikawa men took part in the first fleet sent to the Antarctic in 1934.

Working at the flensing stations provided welcome part-time employment for many in Arikawa during the slack winter seasons, but with the introduction of Antarctic whaling many more people took up whaling as a full-time activity. During the winters (from October through to the end of April) Arikawa men were sent off to the Antarctic, while they spent their summers in the north, either on land stations in the northeast (and before World War II also in Korea, Taiwan and the Kuril Islands), or in pelagic whaling (or salmon fishing) in the North Pacific. Much of the farm land was left to wives and parents to work.

During the heyday of whaling, around 1960, more than 10,000 men were members of the Japanese Antarctic fleets, and of these almost one tenth came from Arikawa. At that time about a quarter of all households in the township had at least one member engaged in whaling. Perhaps not surprisingly, this period also coincided with the population peak for the township, standing at 13,280 in 1960.

Like Taiji, Arikawa has a long tradition of driving schools of dolphins into bays which were then closed off by nets. The town is different from Taiji, however, in that this did not develop into a speciality. Nor has Arikawa turned to small type coastal whaling, and few of its whalers have engaged in this type of whaling so that there are no STCW boats registered in the township. It is also different from Taiji (and Abashiri, Ayukawa and Wadaura) in that there has been no land station in operation there since the 1930s. Despite these facts, Arikawa is very much a "whaling town", although local government employees feel that they cannot advertise the town in this way. "It is too late now", an official admitted. "There is no smell of whales here and only a handful of whalers are left." Although whaling companies have donated artifacts for a whaling museum, it is felt that Arikawa cannot compete with Taiji or Ayukawa in this field, and at present it is uncertain whether the museum project will ever be realized.

Nevertheless, all but the most casual visitor will realize that whaling has been important to this township. A look-out post has been constructed on the site of the old one used to spot whales. Close by is the old memorial stone for whales, put up in 1713. In the centre of the town, too, near the statue of Hara Shinichi, is the Hatsuka-Ebisu shrine where Eguchi Jinzaemon is enshrined. Of more recent origin are the drinking fountain and the yellow switchback formed like cute whales; the dolphin which figures on the town's logo (used on name cards and T-shirts) and telephone cards depicting the old net-whaling method, which are sold at the telephone office and souvenir shops. A restaurant specializing in whale meat is run by a former whaler who worked as a meat inspector for one of the whaling companies, and its walls are adorned with large photos of whaling in the Antarctic. Many people from outside the town come to have a dish of whale meat there.

One of the subjects most often raised by informants in Arikawa is the importance of whale meat for the long history of whaling in Kyūshū has made its mark on the local cuisine with salted fin whale blubber being most prized today. With the high prices at present, people use it with great care. "We eat it as though it were some kind of spice, just to get the flavour of it". Another speciality of Arikawa (and nearby Uonome), which sets these towns apart from others in the Gotō Archipelago, is their liking for dolphin meat. Whenever a school of dolphins is caught, the meat is shared between all the households and subsequently cut into one-centimeter thick slices, before being salted and dried. "All the people used to dry dolphin meat outside their houses, and it was really a pretty sight", one informant recalled. Others made pickles in large containers and boiled dolphin meat is also used in *sukiyaki*.

Whale meat is served at all important celebrations and ceremonies: marriages, coming of age, funerals. On one day in spring, everybody used to have three-layered lunch boxes (*jūbako*) prepared with their favourite food: rice on the lower layer, dishes of whale and dolphin in the middle, and—for children—cakes on the top layer, or—in the case of adults—snacks to go with their *sake*. Even now, about 20 per cent of the annual consumption of whale meat is consumed during New Year and another 15 per cent during "All Souls' Festival" (*obon*) in August (Kalland 1989:112-113).

Whaling has had an important influence on the rituals and beliefs of the population in Arikawa and many religious observances are directly related to whaling activities. These observances, which in the past have given Arikawa its special identity, are being lost under the impact of the moratorium. Morning rituals, for instance, which were performed at home in front of the Shintō and ancestor altars in order to secure good catches and a safe voyage, have been greatly shortened. Monthly pilgrimages made by the whalers' wives during their husbands' absence, are no longer made either, and women today have fewer opportunities than in the past to meet. "Nowdays we only get together during the local water festival and when we have to pay our dues to the housewives' association", one 50 year old woman said. Others felt that village solidarity was breaking down as a result.

The memorial services held for the whales also attract fewer people than in the past, and the temple priest believes this to be the cause of new attitudes, arguing that people have lost some of the respect and sense of interdependence they previously felt toward nature. The village festivals, which center on the shrines, have also changed in character, and the future of the well-known whaling festival, *Meizaiten matsuri* (held on 14 January), is now at stake. Attendance at the more religiously significant *Benzaiten* (held on 12 February) has declined and now attracts only the priest, a few officials and a handful of old ladies. Nowadays, the two festivals that probably carry the most meaning and attract the greatest number of participants are the *17-nichi-matsuri*, which is a festival to avert drowning accidents, and the autumn festival, held during the first two days of

October, which gives Arikawa an identity as a merchant town. Although it has been argued "that cultural activities of whaling can continue long after commercial whaling in the communities has ceased" (The Humane Society of the United States 1987:18) and although Arikawa is specifically cited as an example of this tendency, in fact evidence suggests precisely the opposite.

Conclusion

In this chapter we have introduced six of the most important whaling communities found in Japan today. The descriptions have been sketchy, perhaps, but it has not been our purpose at this stage to give a comprehensive treatment of any one of them. The descriptions are meant to serve as a backdrop for the following chapters, and finer details will be filled in little by little. In particular, we will return to these communities in a more systematic fashion in Chapter 7 in an attempt to analyse what is meant by a "whaling community" in cultural terms.

But already at this stage some salient features stand out. Japanese whaling communities are spread throughout the country, from Hokkaidō in the north to Kyūshū in the south. While Abashiri faces severe winters, with its harbour being closed by ice for a couple of months, Taiji has an almost sub-tropical climate. Moreover, most of the whaling communities are located in remote places. There are few alternative sources of employment, and towns like Oshika (i.e. Ayukawa), Wada, Arikawa and Ukushima have all faced severe depopulation. The remoteness of these communities have also made them likely locations for industries that nobody else wants: a nuclear plant has been built near Ayukawa and this is being enlarged with a new reactor; another nuclear plant was to be built near Taiji, but the plans had to be dropped after strong opposition from local inhabitants; huge oil storage tanks have been built near Arikawa, despite strong local protests. Paradoxically, as whaling is phased out as a result of the activities of international "environmental" organizations, these communities become more dependent on compensation paid by potentially polluting industries. This is one of the many ironies of the present situation.

The whaling communities also vary greatly in scale; from a small enclave within the larger town of Abashiri to Taiji, where practically everybody was once engaged in whaling-related activities. This naturally influences the strategies used by the whalers to defend their interests. In Taiji and Ayukawa, and to a lesser degree in Arikawa, it has been possible to produce a number of new symbols, which all have an important bearing on people's self-identification. This has been difficult to do in Abashiri, a town whose main images are those of drifting ice and the prison. A tourist guide book, which in 1958 described Abashiri as a town of small type whaling (Iwasaki 1987:22), has not managed to alter these images.

There are other differences between the communities described. Some (i.e. Taiji, Arikawa and Ukushima) have been whaling for many centuries, while others (Wadaura, Abashiri and Ayukawa) have been involved in whaling only during the course of this century. It would be a mistake to believe that whaling started from scratch in the latter three communities, however. There are, as we will show in the following chapters, strong continuities between the various whaling communities, both historically, socially and culturally.

CHAPTER 3

Portraits

Having looked at a number of whaling communities in Japan, we will now turn to the whalers themselves, and in this chapter portray the lives of several of those who have been involved in various aspects of Japanese pelagic and coastal whaling. Here our aims are more serious than the mere provision of what some might refer to as "palm tree anthropology". By focussing on the lives of a flenser, crew member, gunner and owner, we want to make plain the variety of career paths open to those involved in Japanese whaling. By bringing in the wives of two whalers, we aim to show how women act to integrate both crew members on a whaling vessel, and these whalers into the community in which they themselves live all year round. At the same time, we are able to provide details of training methods in the whaling industry, as well as of whalers' attitudes towards whales. By revealing how much distress the moratorium on whaling has caused people, and by showing the kinds of problems they have faced in adapting to new ways of life, we want to give the reader an idea of what it means to be a whaler in Japan and to suggest that neither whaling nor whalers are quite so barbaric as is sometimes made out. Here is a group of people who are proud of their skills, proud of the communities of which they are a part. At the same time, however, they share with certain Indian tribes of the Amazon, for example, or coal miners in Britain, the same sense of frustration and anxiety common to all people whose way of life is threatened with extinction.

Abe Yoshio and Noriko: An Ayukawa Whaler and His Wife

Yoshio and Noriko have been married for about 20 years now, but they have come to know each other well only during the last couple of years. Yoshio used to be away for much of the year in LTCW and pelagic whaling. Noriko recalls that she was happy whenever he came back for a visit, "but it was also good to see him go". She used to make all sorts of plans for what they would do together when he returned for a visit. And so did he, but their plans were always different. So after a couple of weeks she was fed up with him and would begin to ask him whether he was not going to leave soon.

And leave he did—every year, until 1988 when he stayed at home. LTCW had come to an end and he found himself laid off. But he was lucky. He got a new job in an STCW company, "although working there nowadays is not like things used to be", he says. "I spend most of my time feeding salmon, because since the moratorium fish breeding has become my employers' main line of business."

Feeding salmon allows Yoshio to live at home much of the year, something which has added a completely new dimension to both his and his wife's lives. "I was a little shy (*hazukashii*) about being with him all the time at first", Noriko smiles. "It was almost like being newly weds again. It's a pity to admit it, I suppose, but that feeling has gone now."

Yoshio and Noriko have managed to make the necessary adjustments in their attitude to each other and to the world, although it has been a painful experience. One problem has been Yoshio's new job. First there is the feeding of the *ginzake* salmon. The fish are handfed from small fishing boats, but whalers are not used to working on such craft. In many countries, autofeeding is the usual way of feeding fish, but in Ayukawa it is prohibited because of the pollution it causes. When the salmon are sent to the market, they have to be caught, scaled, put into boxes, covered with ice and prepared for transportation. Then the workers have to feed the remaining fish so that Yoshio is seldom home before late in the evening during this season. At first he was not used to doing this kind of work, and his hip used to ache terribly. Worse still, his fingers and hands got swollen. "When I woke up in the morning I had trouble moving my fingers." (Yoshio motions with one hand to illustrate his words.) He lost 10 kg in weight and everybody remarked on how thin he had become. Noriko was seriously worried about his health.

Yoshio is still comparatively young, however—only 43 years old—and he has now got used to his new job. He has gained weight again and his fingers do not hurt any more. This has given him back his confidence and he knows that he can manage the kind of work he is made to do. But there are other problems, one of them being their financial state, for Yoshio now earns only a third of his old salary. With two grown-up children, this makes life pretty difficult. The previous year his daughter attended the nearest senior high school in Ishinomaki, some 45 kilometres away, and she had to commute daily because they could not afford to pay board and lodging in the city itself. At the age of 19, she is now attending a sewing school in Sendai, the prefectural capital. It costs at least ¥100,000 a month to keep her in school, and this the parents just cannot afford. So their daughter has taken on some part-time work to make ends meet. Meanwhile, their 14 year old son is in his final year at junior high school and that means that Yoshio and Noriko will be facing another major expense when the boy enters senior high school.

Like most Japanese housewives, Noriko is in charge of both the family's finances and its social affairs. Born and raised in Ayukawa, she has an extensive network of relatives, neighbours and friends to take care of, and the exchanges of gifts in which she is involved in Ayukawa are—to someone unused to rural life in Japan—quite staggering. During the first three months of 1989, for example, she distributed presents to about 20 households, and their value came to a full month's salary. Stuck between the quilt and the formica top of the *kotatsu* table *cum* leg warmer were a number of envelopes—all filled with money—to be delivered during the next few days. Noriko finds it difficult to reduce the amounts. After all, gifts of whale meat have already ceased, so how can she curtail these other presents as well? Gift-giving is ultimately bound up with the Japanese notion of *giri*, or social obligation, and as Noriko firmly exclaims, "we would rather starve than forget our obligations". When Yoshio gives a weak nod, she continues by citing an old local proverb: "You can't forget your underpants and *giri*" (*giri to fundoshi naitomo shinakereba naranai*).

For someone who started out as a housewife, taking care of the house and children, Noriko is a pretty determined and self-sufficient woman. With Yoshio being away most of the time, she was the one who had to make minor repairs to the house, as well as take all major decisions and now she has decided that she ought to earn a living herself. First, she opened up a cake shop in Ayukawa, leaving for Ishinomaki at 4 o'clock in the morning in order to buy fresh provisions. But she got ill and had to close down because of fatigue. Then she turned her hand to running a coffee-bar, which she opened in April 1989. Again, she is faced with long hours because she keeps the shop open until two in the morning. Yoshio helps her after getting home from his own job, and now he has started to lose weight again due to lack of sleep.

The two of them have been through a couple of difficult years. Many of Yoshio's earlier work-mates have left Ayukawa, but Yoshio himself is determined to stay on, because it is only in Ayukawa that he will get another chance to whale. When his old company was dissolved and a new company, Nihon Kinkai, established in its stead, Yoshio was offered a job there, together with another one in an STCW company, Gaibō Hogei. Both had branched out into salmon breeding, but both hoped to continue STCW. "It was very difficult to choose between them", admits Yoshio. "Noriko is a very good friend of the wife of the boss of Nihon Kinkai. On the other hand, I thought the chances of whaling were better in Gaibō, since the company had laid off all its whalers and rehired only a handful of them. Moreover, like me, Gaibō's manager comes from Kyūshū."

Born in 1946 in Hirado, which once figured as an important whaling centre in Japan, Yoshio himself was brought up in a fisherman's family. Still, his mother's brother was bosun on a catcher boat owned by Taiyō Gyogyō, while another relative on his mother's side was chief engineer on one of the same company's catcher boats. Not surprisingly, he heard a lot about whaling from these relatives, and various kinds of whale dish were frequently served at the family's dinner table. He also remembers that, as a child, whale meat was served for festivals, weddings, funerals and so on.

Although Yoshio found himself in touch with whaling in this way from early childhood, he reckons that the incident which came to shape his life more than anything else was when a catcher boat towed a whale into the harbour of his fishing village. He still remembers the excitement he felt then when, as an eleven year old boy, he watched it being flensed. The meat was all laid out on bamboo-racks to dry and, together with some other children, Yoshio later slipped under the racks and cut off pieces of whale meat with some large knives they had brought for the purpose.

It was then that Yoshio decided to become a whaler. This pleased his parents immensely and they later arranged with his mother's relatives to help him find a position. That was the only way you could get a job in those days. "You had to have connections, to know somebody already employed in whaling if you wanted to get a job yourself." So he was taken on by Nihon Hogei, a LTCW company affiliated with Taiyō, and came to Ayukawa, where one of his relatives lived, when he was just sixteen years old. He started as a "boy" before being promoted to deckhand. In Ayukawa he met Noriko, who had been adopted into the Abe household because her own mother was in very poor health. Yoshio took the surname of his wife's adopted family so that he, too, became an Abe.

After getting his seaman's licence (entitling him to act as skipper), he climbed the promotion ladder to become first deckhand. He was promoted to assistant gunner the final year that LTCW was done in Japan. Of course this thrilled him no end, especially when he was allowed to shoot some whales. Noriko, for her part, was perhaps even more excited, and made it her business to go every morning to a local shrine dedicated to the fox deity Inari, taking with her fried bean curd (*abura-age*), raw eggs (since Inari is fond of such delicacies), and sacred rice-wine (*omiki*). During the whole of the whaling season, she missed only two mornings for, like many people in Ayukawa, Noriko has a special fondness for Inari. The deity is believed to protect fishermen and, by extension, whalers, so at their house altar Yoshio and Noriko have three Inari shrines. The oldest used to belong to her grandfather who owned a large set-net and a boat named *Inari-maru* which transported whale meat from Ayukawa to the markets. Then there is the shrine which belonged to her father, and finally a smaller shrine bought by themselves. They also have figures of foxes on their altar. Every morning Noriko used to clap her hands twice and pray to Inari that her husband would catch whales and return safely. Then, when a whale was caught, she offered a piece of whale meat and sacred rice-wine on the altar as a gesture of gratitude, and as a request for more whales the next day.

Every morning Noriko still lights five incence sticks (one for each of her ancestors enshrined there) in front of the *butsudan* ancestral altar and prays that they may rest in peace. She also used to keep them informed about the whales that her husband caught, and both of them still pray regularly for the souls of the whales—that they may rest in peace and be reborn into a higher existence. Memorial services (*kuyō*) for the whales are thus performed daily in the Abe house.

Much of Yoshio's and Noriko's lives is centered on whaling. In his spare time, Yoshio likes to paint scenes of whales and whaling, and he has a collection of pictures painted on baleen. These make popular gifts, although he has now run out of baleen. Their living room is obviously that of a whaler, for figures of whales are found on shelves and a large three-dimensional painting, depicting a factory ship with a catcher boat and whales and signed by Yoshio, hangs on one wall.

The moratorium imposed on Japanese whaling has deprived Yoshio of a golden opportunity to make the envied position of gunner. He has made two trips to the Antarctic and one to the North Pacific, but for the most part worked in coastal waters. In between LTCW seasons he occasionally took employment as deckhand on one of the STCW boats operating from Ayukawa, "just to kill time and get some whale meat". This meant that he was to some extent acquainted with STCW before he joined Gaibō Hogei's boat, and this made it easier for the other crew members to accept him as one of them. Still, he is willing to suffer a lot in order to whale. After all, it is not as if he likes working with salmon, but only those who have agreed to do this work will be allowed on board Gaibō's two whaling boats. Yoshio is thus caught up in a complicated game of "give and take". There is a price to be paid for his remaining a whaler in these difficult times.

But there is also bitterness. It hurts Noriko, in particular, to see the catcher boat *Toshi-maru No. 16* on shore, because this was the boat on which Yoshio worked for seven years. "He knows the boat like the back of his hand," says Noriko, but now the boat has been pulled up out of the sea and put on stilts to become the main attraction for the new "Whale Land" planned by the local town authorities. A lot of people went to see the boat being lifted on shore by huge cranes, but not Noriko. "I would just have cried," she explained sadly. "As far as I'm concerned, beaching that boat was a sign that it is all over with whaling here."

Tokunaga Yasuhiko: A Retired Gunner from Ukushima

Tokunaga is now 65 years old—born in the last year of the Taishō Period (1912-1926), but more or less the whole of his life spent in that of Shōwa (1926-1989). With sunburned face and an enviable forest of black hair brushed back from his comparatively unfurrowed forehead, "Yassan" (as he is called by the local islanders) looks much younger than his years. Now in his second term as *kumiaichō*, head of the local fishing cooperative (FCA), he is a man of standing on the island of Ukushima, in the southwest of Japan. A local fisherman, who otherwise sits silently through a two hour session, during which Yassan talks almost incessantly about his past, clearly regards the *kumiaichō* as indispensable. "We couldn't do without him," he nods enthusiastically. "There's nobody else who can possibly take over from Yassan yet. We'll have to ask him to do a third term of office when elections come round again in a couple of years' time."

Yassan politely demurs: "Well, we'll see what people decide." Then he adds: "Actually, though, I'd like to stop work and do a bit of travelling round Japan with the wife. After all, I've done my stint here. How well, I don't know, but everyone seems happy enough. We're the largest cooperative in Japan, you know, here in Nagasaki Prefecture. And we're pretty big within the prefecture out here on Ukushima. ¥2,600 million turnover last year." He puts on his glasses to check the figures on a sheet of paper that he waves proudly in the air above his desk top. "Mainly abalone. 21 tons of the stuff in two weeks in early May. Going for ¥6,000 a kilo. That's a lot of money, isn't it?"

All the same, abalone is not Yassan's speciality. Having finished schooling at the age of 15, he set out to do what a lot of young men in his day wanted to do: become a whaler. Of course, some kinds of whaling were more prestigious than others. Some did coastal whaling, but it was Antarctic whaling that really counted. So he asked around, begging to be given a job on one of the whaling vessels, and eventually he was taken on as "rice boiler" (*gohan-taki*) on a ship belonging to Taiyō Gyogyō. On that first trip to the Antarctic in 1938, he witnessed a harpooner, Sakamoto, lose the bottom half of his leg when one of the sailors forgot to put a pin in the shackle through which the wire ran to the harpoon. They hauled Sakamoto off to the factory ship, but it was no good. The doctor had to amputate. Whaling, Yassan realized then, was not just glamour.

Then the war came and Yassan found himself drafted off to Kokura in northern Kyūshū. There he eventually underwent training as a marine because that allowed him to go back into whaling (whale meat was an important source of food supply during the war). This he did in 1944, but in the following year things got really difficult, with bombs being dropped and ships being strafed by American planes wherever they went. Yassan was an ordinary seaman then, but was promoted to quartermaster in November 1945. He got his seaman's licence a few years later when he was only 26 years old, and by this time he was sailing regularly on the Taiyō fleet down to the Antarctic.

Although he wanted desperately to become a gunner, he decided that it would be better to work as bosun for a year or two, since that job would give him good training for his future career. (He was right. There were times later on in his life when he was able to take over total control of a catcher boat when it started chasing a whale.) So for three years he stayed up the masthead, shouting instructions down through the voice pipe to the engine room, telling the engineer to speed up, slow down, turn to port or starboard, in order to keep the catcher boat in line with the hunted whale, and so allow the gunner on the platform in front of him to get a good angle from which to shoot. Yassan really suffered in those days. There were times when all he wanted to do was sleep, but he stayed on at the masthead, keeping his eyes peeled for the sight of whales, and showing the others on board that he was really worth his salt before finally learning to be a gunner himself.

It was his capacity for work in fact that eventually allowed him to be taken on as apprentice (*minarai*) to the best gunner of his time, Izui Moriichi, who now lives in Shizuoka. This was quite an achievement, but he had one other vital quality—apart from the fact that he had worked his way up from assistant cook to bosun: he had excellent eyesight. Together with his feel for whaling, this—Yassan reckons—was what made him into such a successful gunner. There are all sorts of people who graduate in this and that and get diplomas at maritime schools, but the really good whalers are those who are prepared to stick things out (*gaman*) and have whaling in their blood (*konjō*).

Anyway, Yassan was lucky. There was a lot of competition in those days and plenty of whales to be killed. Some gunners allowed others to try their hand at shooting with the harpoon gun; others did not. Izui happened to get his finger caught in a door at one stage during their Antarctic trip and couldn't shoot, so he told Yassan to have a go. The young apprentice swallowed hard, concentrated on aiming and was lucky enough to hit a whale first time, right in the middle. Then another gunner on another boat got ill, and Yassan found himself drafted off there as gunner, so that his "apprenticeship" hardly consisted of any teaching as such.

What is it, then, that makes a good harpooner? Yassan reckons that it is "instinct" (*kan*). It is instinct that guides you when you give chase to a whale, and once a whale has dived and come up for air three times, you more or less know its character and can guess where it will surface next time. Still, you have to have a bit of technique, too. Yassan used to move the catcher boat off at an angle once the whale had dived. The sound of the ship's engines would stop the whale from going in that direction and keep it more or less headed straight—especially if you then turned the catcher boat right round and double tracked over to the other side of the whale's path. Then Yassan would bring the boat back to its original position and edge forward slowly. Of course, he was helped by the Asdic (sonic radar machine) that the company bought to help track whales under water, but the operator also had to use his instinct in sending out beams; otherwise, all that technology would be of no use.

The ideal distance between boat and target was about 60 metres, although with larger boats this might come down to as little as 40 metres. At first, Yassan used to aim for the tail of the whale when shooting because, as it plunged into the sea again, that was the part of its body that would hang in the air for a while after its head had gone down. Generally it was "classy" to hit a whale in full flight through the air, but the harpoon used to burn the meat and, as whale meat got in shorter supply, the companies instructed gunners to aim for the head where there was virtually no meat at all.

It always seemed to be so much larger than it really was, the whale, as it reared up out of the water, so—like all gunners—he used to tell himself to "aim high, aim high". If he didn't, the harpoon, which is balanced by the wire attached to it, would fall short in the waves and all that chase would have been for nothing. He had, then, about three seconds in which to sight the whale, aim and fire at it. There was no time to pray to the gods, or anything like that. Still, Yassan always made sure to pay his respects to the "god shelf" (*kamidana*) on the ship's bridge every morning. "Please let me hit a whale, today", he would say and, when he failed to do so, he would occasionally slap the shelf and reproach the deities with a "You must be asleep, or something". Why did he pray? Well, a whale's a living creature, after all, and if you kill a living thing then you should repent to the gods. That was why more than 50 whalers on Ukushima, Yassan included, put up a memorial stone to the whale in the local temple, back in 1952. They should always keep in harmony with nature.

In many respects, a whaling ship was like an army unit. There was great respect for rank. That the gunner was "general" could be seen in the way that everyone would wait for him to arrive at dinner before tucking into their food. Yassan sat at the top of the table, back to the ship's bow, with the captain to his left facing the port side, and the chief engineer to his right facing starboard. Beside them would be the chief communications officer and first seaman. The rest of the crew would eat in the next door mess. Sometimes, if they stopped work early, the crew would cut up raw whale meat for *sashimi*, boil up some entrails and prepare some whale meat for frying, before getting out their duty free liquour and having a party. Usually the engine room crew and the deck hands would mess separately, but Yassan would often join them for a drink because he felt that it was one of the gunner's duties to "mix it" with the crew. Of course, sometimes people would argue when they had had too much to drink, and then Yassan would bring them to order, telling them to drink up and go to bed quietly.

All in all, Yassan spent 36 years as a whaler, before retiring at the age of 52. He could have stayed on until the formal retirement age of 58, but the three largest whaling companies amalgamated and had to cut back on staff. Once he had come home from the Antarctic for the last time, Yassan joined his brothers fishing. Later, he hired a fishing boat for himself, and then—just two years after his retirement—acted as executive officer of the fishery cooperative. Now he leads a comparatively quiet life, devoted no longer to a whaling company operating on the high seas, but to the island village on which he has been born, brought up, and will doubtless one day die.

Kimura Den: A Deceased Gunner's Wife in Abashiri

Kimura Den is the kind of woman who cares. Although she has given birth to and raised five children—two daughters and three sons—she always has the time to listen and help others. She is generous in her love towards people she meets, and she has met many during her 74 year long life. She makes friends easily and is still very active. She made an ideal match for a gunner. "My mother's activities are influenced by her personality, but also by the fact that she was the wife of a gunner", says her youngest son Hideo, himself a devoted whaler, and continues: "A gunner's wife cannot be introverted".

As we have seen, the gunner is the focal point on the catcher boat. In the past it was he who decided who should crew, and even today some company owners, particularly in STCW, leave this aspect of recruitment to the discretion of their gunners, since they know that if a crew is not dedicated to its gunner, the boat will not do well. Still, it is often the wives who have the most widely spread social network in local communities all over Japan, and this is equally true of the Kimura household. It was always Den who had to send whale meat back to her relatives in Yoichi, where she was born, as well as to Arikawa, where her husband came from. She did so until she ran out of supplies after the moratorium was enforced.

Den has brought several young men into whaling. In the past it was always a pleasure to do friends a favour by helping their sons onto a catcher boat, but now she is overcome by feelings of guilt. "You realize that now they have become unemployed." Her otherwise smiling face is serious. "Take one of our neighbours, for example. His mother asked me to introduce him to my husband. That was in 1973 and now he's unemployed. His wife doesn't work because she has a small child to take care of, as well as his old and infirm father. His savings have all been spent."

Den has always been concerned about the welfare of her husband's crew. Her home has always been open to the whalers and in some respects she is like a substitute mother for them. In the 1950s it all became a bit too much for her, however. Her husband, Masao, was working for an STCW company called Hōyō Hogei at the time. They lived in a company house and their home was open not only to whalers employed by Hōyō, but to Arikawa whalers employed by Nissui as well. In those days Nissui had a flensing station in Abashiri, and many whalers were brought from southern Japan. Masao was himself born in Arikawa, and Nissui and Hōyō were affiliated companies so it was difficult to turn the Arikawa whalers away. "They brought their bottles of drink with them, but I had to prepare all the food to show our good will." In order to get a modicum of privacy, and to protect their five children from the noise which these visits frequently brought, the family decided to move out of the company house.

Den was born in 1915 at Yoichi, not far from the large fishing port of Otaru in Hokkaidō. When she was 23 years old she met Masao at Hakodate, where Masao's boat was docked and they got married shortly afterwards. She followed him around, first to Kiritappu and then for four months in Chishima and Shakotan north of Hokkaidō, before later accompanying him through the Inland Sea to Korea. "I lived on board the whaling boat while we travelled from place to place", she recalls. She was not the only wife there, for both the captain and chief engineer brought their wives with them too. "We slept in shifts because there weren't enough cabins. Then when we got to the whaling grounds, we rented houses. We were never allowed to stay on board during the hunt. That was taboo!" Women are still regarded as unclean and thus polluting to the boats.

This nomadic life lasted for less than two years before the Pacific War brought it to an end. Life on the catcher boats became too dangerous, and Den was sent to Mogushi in Arikawa, to the hamlet where Masao had been born in 1913. A house was built for the young bride, and she tried to adjust to the unfamiliar surroundings in the south. "Not much happened, really. Mogushi was a very silent place." Den tries to recall the past. What does she remember best? "Surely the dolphins. The only time Mogushi came to life was when schools of dolphins were landed on the beach. The whole place was covered with blood. We salted the meat and ate it for dinner, although some people ate it raw."

Life could be lonely for her in Mogushi since she did not know many people there and Masao was away much of the time. Although the old Arikawa shore station had been closed a couple of years before Masao was born, Mogushi and a number of other hamlets in Arikawa lived off whaling activities. Her husband had been surrounded by whalers from birth; it was an uncle who got him, and his two brothers, into whaling. He was 16 years old when he was first hired on a catcher boat, belonging to the biggest Japanese whaling company at that time, Tōyō Hogei. He was sent off to various shore stations and managed three trips to the Antarctic before the outbreak of the war. He was then bosun for Nissui, the company which absorbed Tōyō Hogei in the mid-30s.

During the war it was impossible to continue pelagic whaling, and LTCW was also greatly affected since most of the whaling boats were taken over by the authorities. At the same time, however, there was a shortage of food in Japan and coastal whaling received special consideration. Nissui bought five STCW boats, and Masao got employed on one of these. Shortly after the war, he was sent to the South Pacific for a season, but decided then to go back to STCW. He suffered from asthma and preferred to work in STCW so that he could return home almost every day. He got employment with Hōyō Hogei, a Nissui subsidiary that had taken over some of the latter company's STCW boats after the war. The company operated out of Abashiri, so in 1946 Masao took his wife and three children north, first to Yoichi, Den's home town, and a year later to Abashiri where the two youngest children were subsequently born.

Like many whalers, Masao was restless by nature and tried to move on. He would have been successful but for some behind-the-scenes intrigue which prevented him from taking over an STCW boat in Ayukawa and thus becoming an operator himself. When his youngest son was only four years old, Masao moved his family to Hayase, a village on the Japan Sea coast way down in Fukui Prefecture. There he started to hunt minke whales and in the off-season tried to make a living by fishing for sea bream (*tai-ippontsuri*). Still, his attempts to catch whales in Wakasa Bay were not very successful and after a couple of years he moved back to Abashiri.

As gunner, Masao was in a position to take his children on board. "All my sons loved to fish, and they all accompanied their father on the boat", says Den. Sure enough, the youngest son can still vividly remember the first trip he took when he was only eight years old. "It was a beautiful day, and my father said there was a good chance of catching a whale. We caught two or three, and I thought it must be terribly easy to catch whales. How disappointed I was when I later discovered that it wasn't quite so easy after all!" Hideo laughs. "You know, strangers aren't allowed on board, but children are. Some children bring good luck, others bad." His eldest brother apparently brought bad luck, for even though he went out on the boat several times, they never caught anything and the crew used to joke a lot about it. Whether this influenced him against becoming a whaler, though, is doubtful. According to Den, he was such a good baseball-player that he had no difficulty in getting employment elsewhere. He has taken over his father-in-law's business, but he often pops into the sports equipment store that his mother now runs, together with her daughters-in-law, located along the town's main shopping street. Now that his two brothers are unemployed, they have nothing to do but hang around or kill time by giving a hand occasionally, even though their help is not really required.

Masao instilled in all his children a high regard for whales and whaling. "He taught us everything we know", says Den, before adding with a laugh: "He was so superstitious! Even when he dropped a knife at home, he had to purify it with salt, the way they always do on the boat!" She was deeply influenced by her husband's religiosity and learnt, like him, to take both deities and ancestors very seriously. Every time a whale was caught, a piece of the tail and some meat were offered to the boat's "god shelf" (*kamidana*). Masao used to take some small plates from the kitchen to place the meat on, but they always got broken on board and Den used to get angry with him for stealing them from home. He also brought meat home, and placed some of it on the "god shelf" there.

But Den was also active in placating the deities and ancestors. When her husband was out whaling, she used to go to the local shrine for the fox deity to ask for help, or to give her thanks. Moreover, she prayed every day for good catches and a safe voyage in front of the family's ancestral altar, the *butsudan*. This is the first thing in a house that should be saved in case of fire, and indeed, when Den's house did burn down three years ago, the only thing they managed to get out was the altar. Ironically enough, this happened early on New Year's morning, but in fact Den had been over at her youngest son's house the evening before and had taken Masao's memorial tablet (*ihai*) with her so that they could celebrate the coming of the New Year together. Masao is still with her always.

Nakamura Kanji: A Craftsman in Ayukawa

Nakamura Kanji was born in Karatsu, an old whaling centre in Kyūshū, in 1928. As a child he was adopted by his uncle on his father's side, Nakamura Seiichi. His real father, Okada Asao, had inherited the family business from his father, making plectrums for the Japanese *shamisen* lute out of sperm whale jaws. But supplies of this material were running short in southern Japan, and it was realized that somebody had to go out to seek new supplies elsewhere. Seiichi was a stone mason who specialized in making sculptures, and as it was argued that he could more easily find a job in northern Japan than his brother-in-law Asao, it was decided that Seiichi should make the then dangerous trip to Hokkaidō, where there were apparently rich supplies of whale jaws.

Seiichi never got that far, however. In Sendai, he found out that Ayukawa was in all likelihood the best place to look. Sure enough, the prospects of doing good business in Ayukawa were so bright that he decided to settle there, and after a couple of years he brought his family—including his adopted son—to the town. Kanji was about five years old then. He thus grew up in Ayukawa and has spent most of his life there. When he was young, he watched his step-father buying whale jaws in order to send them to his real father, but then Seiichi gradually stopped doing this and decided to increase his own production instead. "I used to hang around watching my step-father turn the bones and teeth into various small objects", Kanji recalls. "Being adept at making small stone sculptures, he found it comparatively easy to work with the whale bones". The crafts that he made, he soon started to sell through a friend of his who was a captain on one of the Nissui boats.

By the end of the war Kanji was helping his father in the workshop. Still, they faced serious problems. Most of their products had until then been sold in northern Korea through the Nissui captain, but with the loss of the Korean market they had to built up a new network of customers in Japan. There were other difficulties, too. During the war it was prohibited to make ornaments from the bones and teeth, so that their products had consisted of walking sticks, cigarette holders and family seals (*hankō*). But the demand for white sticks fell away drastically when Japan decided to follow international usage in having white walking sticks to signal blindness. Later, the market for cigarette holders collapsed when the filter cigarette was invented, so the Nakamuras found themselves having to learn to make new ornaments.

In 1952, when Kanji was only 24 years old, Seiichi died. He can still remember what his step-father said just before he passed away: "You're well trained and you can continue alone!" But Kanji has not found it easy to stay in business, for the market has changed considerably over the years. At one stage, most of the customers were whalers and seamen, who wanted to buy whale products which made nice gifts, but with the decline of whaling and shipping the number of such customers naturally declined. On the other hand, the opening of the "Cobalt Line" scenic drive has brought many tourists to Ayukawa—tourists who have very different tastes and preferences from those of the whalers and sailors whom he had hitherto served. Beside his own shop, Nakamura now sells his products through tourist outlets: the ferry terminal store, souvenir shops on Kinkazan, and many of the town's hotels. "It took about ten years to adjust to this change in market", he says. "And by the time we finally managed to change our product line and build up a new image, whaling had become doomed."

He is naturally anxious about the future—as is his second son, Tetsuo, who plans to take over the shop. After completing university, Tetsuo was sent by his father as an apprentice to a craftsman in Kawasaki City, outside Tōkyō, where he spent four years learning how to make family seals. "It is very difficult to train your own son", is one reason Nakamura gives for sending his son away. Another reason was that the teacher in Kawasaki worked in a number of different raw materials, and Kanji thought it useful that his son should be able to work materials other than whale teeth. He himself does not think that he can make the switch to new materials. For a start, he does not have the energy to go through a long new process of adjustment and renewed image building. He is fortunate, therefore, that he still has some whale teeth and that he has been very careful not to waste anything in the past.

Seiichi's moving to Ayukawa around 1930 and taking up a whale handicraft has influenced some local people who have also tried to learn the necessary skills. Many whalers found carving teeth an interesting way to kill time while sailing to and from the Antarctic and carved teeth made excellent souvenirs. Few of them, however, ever became professionals, although one exception is his own brother, Okada Akira. On the advice of Seiichi, Akira moved to Ayukawa after the death of his father and learned the trade from his uncle. Anothercompetitor is Endō Junichi, son of an Ayukawa flenser from Arikawa, who started accompanying his father on a factory ship to the Antarctic. He found it more interesting, however, to sit in the shade carving whale teeth than to flense whales and when his talent was recognized by Nissui, he was allowed to continue his craft, provided that he handed over all his products to the company. Junichi managed to establish himself in business, and it was he who took the initiative in approaching the Norwegian Embassy in Tōkyō to ask to be allowed to buy sperm whale teeth from Norway. Unfortunately for him, the request was not granted, even though Junichi later learned that several sperm whale carcasses had at that time drifted ashore in Northern Norway, causing pollution and inconvenience to the local fishermen. Some of these carcasses had then been towed out into the open sea and destroyed by dynamite.

It was Kanji's parents-in-law who told him to put an image of the whale on the *kamidana*. Business would go well, they said, if he placed a whale image there. Every morning he bows in front of this "god shelf" and prays to the village tutelary deity, as well as to the memory of the whales. "When I pray to the whales, it is mostly in gratitude to them. It's thanks to them that we can make a living. I also always attend the memorial service held by the township for the whales in connection with "All Souls' Festival". All those who are directly dependent on whales and whaling attend".

Whales have provided Nakamura and his colleagues with work and an income. Today they are in difficulty. Relief was only temporary when the whaling companies decided that professional craftsmen like Nakamura should be allowed to buy the last teeth they had, thus ignoring the pleas from amateur craftsmen *cum* whalers. But their final supplies are now running out and they feel forgotten. The town is more concerned with unemployed whalers than with underemployed craftsmen. "The whalers have got a lot of compensation but we haven't got a single yen", Nakamura complains. "Nobody thinks about us at all."

Shimoda Tamekichi: A Retired Flenser from Arikawa

Shimoda Tamekichi likes to entertain. Glad to have the opportunity to talk to foreigners about whaling now that he has little else to do, he pulls out a couple of cushions and calls to his wife to prepare some tea. He was born, he says, in 1911. His father worked for Tōyō Hogei as a cook, and was sent off to work at a number of different shore stations over the years. When Tamekichi graduated from school at the age of 16 (there seemed to be no point in continuing with education in those days), it seemed inevitable that he should start working for the same company as his father, and he found himself being sent first to Tsushima and then to Hokkaidō. Being a long way from home and at a different shore station from the one on which his father was working, Tamekichi felt lonely in the beginning. He remembers how he cried a lot in his bed at night. But gradually, in spite of his tender years, he got used to his new role and came to enjoy life.

The first year he worked as a "boy", running errands in and about the office, but then he became a cook, a job he held for two years. Only after that was he allowed to do "real work" with the whales (*kujira-shigoto*)—first as a *kagihiki* "puller" who helped the flensers make nice, firm cuts by quickly pulling meat and blubber from the carcass of the whale. Then, after another two or three years of this work, he was allowed to handle the flensing knives. He felt really elated then because, as he now recalls: "Only four or five out of a hundred ever made it to become flensers".

It takes several years to become a good flenser. At first Tamekichi did not get any formal training, even though a relative on his mother's side was an experienced flenser at the same shore station in Ayukawa. This he really regrets. "People in Arikawa where I came from wanted to keep their skills for themselves", he explains sadly, so he had to learn the hard way, observing what other flensers did, before finally finding a real teacher—a local flenser from Ayukawa called Furukawa. "He was really marvellous at flensing sperm whale, and he taught me a lot. To me, he was like a god (*kamisama*). It was only after Furukawa taught me how to cut sperm whale that I reckon I became a good flenser".

Tamekichi is by now well into his stride, and decides to celebrate the anthropologist's visit with some *sake* and a little of the frozen whale meat that he has in the freezer in the covered passage at the side of his three room house. He brings in an electric cooking ring and places it on the low table at which we are sitting, before pouring in soy sauce and sugar to prepare for a whale meat *sukiyaki*. Before the war, he continues, he used to work in the north of Japan during the summer seasons. For several years he more or less commuted between the shore stations of Shakotan in Hokkaidō, and Shana and Toshimo on Etorofu. During the winter seasons he went off abroad to work at stations in Taiwan, Korea and Ogasawara. Like many of his friends, he wanted to join the Antarctic fleets of his new company, Nihon Suisan (Nissui), but his father refused to give him permission to go—permission that was necessary in those days.

Tamekichi was only 35 years old when he became master flenser (*i'inchō*). He started out as before on shore stations, but after the war his company asked him to go to the Antarctic, and he was not one to refuse such an offer. So he finally crossed the Equator for the first time at the ripe old age of 40. Although on land he had been appointed supervisor, at sea he found himself being treated as a youngster again, required to do such menial tasks as preparing food and working as a *kagihiki*. It was a humilating experience, especially as other younger workers on board the factory ship did not think much of him. "You see, there I was—a middle aged man almost—working as a *kagihiki* and preparing food, so they all thought that I was a good for nothing, and that I had been unable to make a career for myself. But I soon showed them. They got quite a shock when I was appointed as master flenser again the following year!"

One of Tamekichi's most important duties in this job was to keep records on all the workers under him. This meant that he had continually to evaluate them and give them marks on a scale of one to six for their performance on the flensing deck. "It was a difficult job," he sighs as he refills our *sake* cups, "But the main thing was to mark each worker according to how sincerely he worked." The real problem was that these marks influenced the salaries the workers received, too, so it could put Tamekichi in a really difficult position vis-a-vis his colleagues at work. Not surprisingly perhaps, this system was eventually abandoned as a result of strong union opposition.

Possibly one of the most interesting experiences of his life came when he was asked to act as observer aboard a Russian whaling vessel, and Tamekichi fondly recalls the sense of camaraderie he felt with other whalers. They could not communicate that well, of course, since they all had to speak very broken English, but they downed quite a few bottles of vodka and that was fun. The Russians wasted a lot of the whales that they caught, though, and that was a pity. In this respect, he feels that the Japanese whalers were really efficient. After all, they have always made use of the whole whale and did not throw large quantities of it back into the sea.

Tamekichi continued to work as master flenser until 1966, when he reached the customary age of retirement in Japan, 55 years old. Nissui still needed him, however, and asked him to stay on as warden at a dormitory for young Nissui factory workers at Hachiōji, near Tōkyō. This he did until he retired properly once and for all in 1976. It was then that he was finally able to settle down and make friends in his home town. "I'd been away so long, I felt as if didn't have all that much contact with people in Arikawa. So I took up gateball. You know, that game all the old people play around here? It's quite good fun really, and I've got to know quite a few of those who live around here. Of course, most of them were whalers in their time and we often talk about the good old days in the Antarctic or on the shore stations." He deftly picks a piece of slightly sugary whale blubber from the steaming pan and dips it in the raw egg in the bowl in front of him. "It's sad, though. Really sad. There's nobody now who can flense right whales or blue whales. Soon, there'll be nobody left who can flense at all. And that'll be the end of it."

Sasaki Reisuke: A Boat Owner

Sasaki Reisuke is the owner of an STCW whaling vessel in Ayukawa. Now in his mid-70s, his age is belied by the energy with which he lives his everyday life—whether it be attending the flensing of a Baird's beaked whale in the little shed in front of the town's main hotel, bargaining with local wholesalers over the price of whale meat, explaining what whaling means to people in the local community to visiting newspaper reporters and anthropologists, or pursuing the cause of small type coastal whaling both at home in the Ministry of Fisheries, and abroad at an IWC meeting in somewhere like San Diego. He is able to convey his arguments cogently and with a seemingly unlimited enthusiasm, and those who meet him regularly are convinced that he is getting younger with every year that goes by.

Reisuke was born in Iwate prefecture in the north of Japan, and first came to Ayukawa in 1933 when he was 18 years old. In those days, quite a few young men from the neighbouring villages used to come as a group and work on the nets off the island of Kinkazan. Reisuke's father had a couple of bonito fishing boats, on which he employed about 20 men, but he was not very good at business and the family's financial situation was not so good, so Reisuke himself decided to leave his home village. He was second son, and did not have to worry about taking over his family's affairs. He ended up joining twenty or so other young men who were recruited every year to work on the Kinkazan nets. Two or three other young men from his village had been down to Ayukawa, and Reisuke knew how lively and prosperous the town was and how much money was to be made from whaling. The first year he spent working on the nets, in his spare time looking for an opportunity to get on board a whaling vessel, but he did not have the necessary contacts then, so instead he spent the second year crewing on a fish transport vessel, just to show the locals how hard he could work.

His diligence paid off when he eventually found employment as a third deckhand on board the local whaling company boat, the *Ayukawa-maru*, a ship of almost 150 tons. It was a pretty tough job, though, and Reisuke found himself obliged to be on watch two nights out of three right through the whaling season. A lot of the time this meant rowing crew members back and forth between the whaling ship and the harbour, where those off duty used to drink and frolic to all hours of the night. When they wanted to leave for the ship, one of them would blow on a whistle and Reisuke then found himself in his rowing boat once more. Quite often, too, the captain would decide to set sail in the middle of the night and Reisuke had to rush about the whole village rounding up members of the crew who, by late evening, were well into their cups and in no fit state to go anywhere much. It was really hard work, he sighs (and his wife, who is from Ayukawa, adds with a chuckle: "Hard work because he wasn't out drinking himself!").

He acted as general dogsbody in this way for a year and a half before being promoted to first deckhand—a job that he held onto for another 18 months, when he started acting as apprentice gunner. The *Ayukawa-maru* had been taken over by a new whaling company which wanted to send a fleet to the Antarctic, and he and his fellow crew members were trained for new jobs because of the shortage of crews for the catcher boats that were to head south of the Equator. After a year's training, Reisuke found himself being sent to the Antarctic in 1939.

Then the war came and he was eventually drafted and sent off to China. Returning to Ayukawa in 1947, he found that almost all the whaling vessels had been destroyed, but his old company re-employed him and sent him off as gunner on a catcher boat hunting whales in the waters off Hokkaidō. His eye was good and he shot more than twice as much as any other gunner, so that things were looking good until 1949 when an old friend of his (in fact, the man who had owned the Kinkazan nets on which he had worked when he first came to Ayukawa) asked him whether he would not join him in whaling. His friend had a boat, but no crew. After one or two misunderstandings, Reisuke gave up his extremely well paid job with the whaling company (where he was earning ¥24,000 a month) to become gunner on an STCW vessel. His pay there was only ¥7,000 a month, but he and his friend (who had put up ¥5 million for the boat) agreed to split the profits between them. Their expenses were high—as much as ¥3.5 million a year—but they were more than offset by an annual turnover that reached as much as ¥10 million a year. In 1950, when they started their joint venture, that was a lot of money.

After several years, Reisuke's friend, who did not actually work on the whaling boat, went bankrupt and Reisuke soon found himself acting the combined roles of gunner, captain and owner. When he reached the age of 50, however, he stopped shooting whales and he has since then concentrated his efforts on running an STCW business in Ayukawa, as well as venturing into the volatile tourist trade. In recent years, this has meant that he has had to be politically active—both in local government, where Reisuke has in the past objected to shady transactions, and in national and international whaling affairs. As a result, he has found himself travelling frequently between Ayukawa and the Ministry of Fisheries in Tōkyō, where he has put the case for small type coastal whaling and listened to seemingly endless hours of bureaucratic stone-walling. He has also been to one or two of the International Whaling Commission's annual meetings, ostensibly to put the case for STCW to the world, but he has at the same time had to witness what appears to him to have been a total about-face by Japanese officials when up against international opposition to their country's whaling. In his old age, he is not afraid to refer to them as "crafty" (*zurui*). But then, that is what government in Japan is all about. Nobody is prepared to take responsibility for anything these days.

CHAPTER 4

The History of Japanese Whaling

Three forms of whaling existed in Japan when the whaling moratorium went into effect in 1987: Pelagic whaling, LTCW (large type coastal whaling) and STCW (small type coastal whaling). Of these three types, pelagic whaling is the most recent phenomenon in Japan, but the other two types have been carried out there for many centuries—even millenia—during which time there have been many changes in technology, in species caught, in hunting grounds, and in the location of the whaling communities themselves. It is the purpose of this chapter to outline the historical processes leading up to the three modern forms of whaling which existed when the moratorium was enforced. In so doing, we will describe the development of each of these three forms, in order to illuminate the historical continuities in Japanese whaling. At the same time, we will describe the social organization of pre-modern whaling in order to provide background material for an understanding of the continuities which, we will argue in the next chapter, exist in the ways whaling activities were carried out until 1987.

Five Stages of Whaling

When Fukumoto Kazuo (1978) suggested that the development of Japanese whaling can be divided into five stages, he first of all had LTCW in mind. According to Fukumoto, the first of these stages lasted well into the sixteenth century when whaling was not yet established as a business. Although whales were occasionally hunted by bow and arrow and even nets, it was mostly dead or wounded whales which were caught as they drifted by. This kind of whaling has been called "passive whaling", as contrasted to "active whaling" where fishermen pursued healthy animals (Hidemura and Fujimoto 1978:163).[1]

Active whaling is thought to have started in the sixteenth century, but it was only towards the end of that century that whaling developed into a large-scale enterprise, thereby marking Fukumoto's second stage. Here whalers rode in several boats and made use of harpoons in the hunt, a technique that come to be known as the harpoon method (*tsukitori-hō*). Killed whales were brought back to specially established processing facilities on shore. This technique was particularly important in Wakayama, Shikoku, Northern Kyūshū, and on the coast of Yamaguchi facing the Sea of Japan.

Frame 1: "Passive Whaling" in Tokugawa Japan

The emergence of active whaling around 1600 does not mean that passive whaling ceased to be of importance. On the contrary, the commencement of active whaling increased the number of dead and wounded whales that drifted in the sea. The spotting of such a whale was always accompanied by the greatest excitement among people—so much so that fights sometimes erupted between villages over who was entitled to the whale. Although most fishing villages in Japan have exclusive fishing territories within which the fishermen can lay claims on all resources, territorial lines are not always easy to draw at sea. Moreover estuaries, which in the old days were often used to demarcate village territories, occasionally changed their course.

In one such case, in 1829, a whale drifted to what Eguchi Village (in Fukuoka Prefecture) claimed was their territory (*Akashi Documents* No.258 and No.331). The village tied up the carcass and set up guards to prevent theft of whale meat, a practice which was common in those days. However this was discovered by the village to the east, Kanezaki, which sent "hundreds" of people, many "with knives to cut up the whale, and there was a fight between the fishermen of Kanezaki and the guards of Eguchi". Things got worse when the village to the west of Eguchi, Kōnominato, came to Eguchi's rescue—so bad in fact, that

---

[1] These stages overlapped considerably. Passive whaling has always been conducted, and a stranded whale was recently flensed and sold through the markets in Hokkaidō (Bestor 1989b). Whaling with harpoons continued long after the invention of the net method in the 1670s because it was much cheaper and easier to organize.

Eguchi's village headman "thought that if it continued like this until it became dark, it would grow even worse. So I recalled the guards, and the ropes tying up the whale were cut." Eguchi was apparently able to retrieve this whale again later, as it was sold in an auction in the village a few days later (*Akashi Document* No.332).

In those days, two thirds of the price of a whale was paid in tax to the feudal authorities in Fukuoka, who, however, put the money into a special development bank for fisheries. The remaining one third was distributed throughout the village (Kalland 1988:204). In spite of the fact that the village only retained one third of the income raised, it was still worth people's while to go out to retrieve drifting whales.

The villagers of Genkaishima (also in Fukuoka Prefecture) were in the middle of their New Year's celebration when a drifting right whale was spotted on the second day of the year 1861. A contemporary document, *Kenbun Ryakki* ("Notes on things seen and heard"), describes the incident: "As the sun would soon set, the timing was bad. But it was hard to give up just like that. All the young men on the island quickly made their preparations and four boats were made ready with such speed that there were matters which were not properly attended to. Everybody jumped on board. It was now near sunset, when the sun takes on colour ... northeastern winds became stronger at nightfall and built up high waves ... They tried to return home ..., and three boats managed with difficulty to do so. However, one boat with nine persons on board did not return that night." The wrecked boat and two corpses were found on the 6th. Meanwhile, on the 4th, another village had retrieved the whale, which had, though fatally wounded, escaped one of the net-groups from Iki Island.

Many of the whales retrieved were reported to be "only skin, bones and entrails". One reason for this situation was that the whales often drifted in the busy shipping lanes and "it seems that (people on these boats) little by little cut away (the meat)." One whale retrieved by Nishinoura in 1826 was found to have only 10 per cent of its meat, 30 per cent of its entrails and 40 per cent of its baleen intact. "What little that remained was moreover completely rotten", says the report. Nonetheless, four flensers called in from another village worked half a day on the carcass, and they "were strictly guarded. How it would harm ourselves if it was found out later that people of this village had stolen and hidden meat!"

Theft of meat was a problem both during passive and active whaling. Illustrations of flensing stations usually depict guards chasing people away, and an eye-witness from Tosa (Shikoku) stated: "We ...

watched the fishermen carve up the whale in the sea. As they did,
scores of local housewives jumped into the water and tried to cut meat
from the dismembered carcass. Angry fishermen brandished bamboo
poles to drive away the women, who did not flinch. They were like
flies on the back of an ox. Some were beaten and injured about the
head. It was an interesting sight." (Statler 1984:233.)

The Net Whaling Method

Toward the end of the seventeenth century, Japanese whaling
entered its third stage through the invention of the net method
(*amitori-hō*), and preliminary trials were at least made in Taiji and
Kayoi (in Yamaguchi Prefecture). The technique developed in Kayoi
was the simpler of the two. Whales were pursued into a narrow bay
whereupon a net (*tachikiri-ami*) was set across the entrance, blocking
escape. Several villages on the Nagato Coast had by the 1670s started
to use this method (Tokumi 1971).

It was, however, the method developed in 1675 by Wada
Kakuemon in Taiji which was the more efficient of the two methods.
Here, large groups of people were organized to drive whales (mainly
slow moving species like right and humpback whales) into large nets
(*tenraku-ami*) set in the open sea. As soon as the whales were
entangled in the nets, they were attacked with harpoons.

This method spread rapidly throughout most of southwestern
Japan and continued to dominate Japanese whaling until the end of the
nineteenth century. This kind of net whaling involved substantial
capital (including investment by Ōsaka merchants) and the net
operators themselves frequently moved from one whaling ground to
another, taking skilled workers with them. Rights to exploit particular
whaling grounds were granted by feudal fiefs (*han*) in exchange for
fees to the authorities and compensation to local communities for the
inconveniences whaling operations caused them. Moreover, regulations
were often extended to the distribution of meat in the form of payment
to the whalers and compensation to the villages affected by whaling
operations (Fujimoto et al. 1984).

Given that the organization of these net groups was in many ways
similar to the organization of modern whaling, we will briefly outline
here the main features of the net method. The operation fell into three

separate stages—preparations for a new season, hunting, and processing the whale—with each of these stages requiring special skills and modes of organization.

The preparations for a new season (*maesaku*) usually started in September and included a range of activities centered on the shore station (*nayaba*). First, large quantities of hemp were brought in, often from a considerable distance, as raw material for ropes to be made by women living in the host and neighbouring villages. Then male experts (*ami-daiku*) used the ropes to make new nets, since many of the old nets had to be replaced every year. These experts were usually recruited from villages that specialized in this trade, and often from distant provinces. Every year some of the oldest boats had to be replaced by new ones built by specially employed boat builders (*funa-daiku*), while harpoons, knives, containers, and so on were made by smiths and coopers. Moreover, the working sheds had to be repaired or rebuilt, new furnaces made and firewood collected.[2]

Next, several activities were involved in the hunting of whales. First, once the weather was regarded as suitable for the whaling season to begin, the search was initiated. Lookout posts on hilltops, commonly manned by five persons, sent smoke or flag signals to convey information to the shore station about whales that they had spotted. In areas where there were no suitable lookout points, search boats were used to look for whales and the shore station was informed of sightings by means of flag signals.

Once a whale had been spotted and the shore station was informed, between ten and twenty swift hunting boats (*seko-bune* or *oi-bune*), each carrying a crew of about twelve persons under the command of an expert harpooner (*hazashi*), set out in pursuit of the whale. The boats split into three groups each led by a chief harpooner (*oyaji*). By surrounding the whale on three sides and by beating the sides of the boats, they managed to frighten the whale in the desired direction. Meanwhile the net-boats (*sōkaisen*) and their assistant boats (*amitsuke-bune*) arrived on the scene and the nets were lowered under the direction of a commander-in-chief (*mito-oyaji*) through signals to

---

[2] 8,165 working days were spent during this stage by one net group consisting of 61 identified artisans plus an unspecified number of firewood cutters and female rope makers at Tsushima in 1802 (see Takeno [1979] or Kalland [1986:34] for details).

the net-boats which worked in pairs—one pair for each of the nets.[3]

As soon as the whale was entangled in the nets and its speed had been slowed down, the hunting boats approached the animal and the harpooners threw harpoons secured with ropes toward the whale. The first harpooner who managed to spear the whale was financially rewarded. The most daring task was accomplished by one harpooner who had to climb onto the whale's back, cut a hole near the mammal's blowhole, and thread a rope through this hole to secure the whale. Another daring operation was to dive under the whale with ropes and tie the whale to two beams laid between two boats that served as floats (*mossō-bune*). Only after this had been done was the animal killed by a sword and the whale towed to the shore station by the *mossō-bune*, leaving the hunting boats to chase other whales that might have been spotted in the area.

Whale processing was carried out at shore stations containing a number of working sheds (known as *naya*), living quarters, offices, and winches—all centred around the beach up which the whale was dragged. The shore stations varied somewhat in their physical lay-out, but the different stages in the processing of the whales were mostly the same. These involved, firstly, bringing the whale from the sea onto shore by means of hand-powered winches (*rokuro*) which were then used to strip the blubber from the whale as the main flensing (*uo-kiri*) began. This *uo-kiri* (lit., "fish cutting") consisted of rough hewing of the whale's carcass and simultaneous separation of the blubber from the meat. This was followed by "middle cutting" (*naka-kiri*), in which the meat and blubber were cut into smaller pieces capable of being carried by two men with a pole. Because of the special requirements of Japanese cuisine, both *uo-kiri* and *naka-kiri* were carried out by highly skilled flensers.

All of these activities were conducted out in the open, before the meat and blubber were carried indoors. Although each shore station varied to some extent in its organization, we find that there was a frequent separation of sheds where meat, blubber or entrails were further cut into smaller pieces and processed independently. The major part of the meat was used fresh or salted as food. The blubber was mainly boiled and used to extract oil, which was in great demand as insecticide (Fujimoto 1967). The entrails, for their part, were used

---

[3] Usually there were three nets set outside each other, but some smaller groups had only two nets.

both as food and for oil production. The whale's bones were taken to a separate shed where they were crushed and processed into oil or fertilizer. In many *nayaba*, sinews were also processed in a separate shed, as were sperm whale teeth and baleen. Processed sinew was used in a wide range of products such as musical instruments and bow strings; sperm whale teeth and baleen were also utilized in a variety of crafts, including the making of *bunraku* puppets; *shamisen* plectrums were produced from whale jaw bones. There was thus virtually total use of the whale. This was in sharp contrast to the western usage of whales at that time.

The American Method

In 1820 the rich whaling grounds between Hawaii and Japan were discovered by American and British whalers and within a few years hundreds of whaling ships from the United States and other western powers were operating in Japanese waters. In 1846 there were almost 300 ships from the United States alone (Tønnessen 1967:178).

Not surprisingly, perhaps, the activities of western whalers coincided with—and are widely believed to have caused—a drastic reduction in the number of whales caught by the Japanese in their nets, for the whales were being caught before they reached coastal waters and came within reach of the small Japanese rowing boats. For example, the catches registered by the Katsumoto/Maeme net-group on Iki Island went down from 138 whales in 1845 to only 14 in 1856 (Akimichi et al. 1988:12). There are many similar complaints about the reduced catches of whales, and many of the net-groups fell into serious financial difficulties. It was obvious that if Japan wanted to remain a whaling nation, she would have to modernize her operations.

The Japanese turned to American models when first attempting to update their whaling industry. After all, the Americans had in the early nineteenth century replaced the Dutch and the British as the leading whalers in the world. Originally catching right whales from shore stations in New England, from about 1760 the Americans started to catch sperm whales further out to sea as a result of declining catches of right whales. At first they went to Davis Strait, Baffin Bay and other Arctic waters, but within a decade whaling ships from New England had started catching in the southern Atlantic. The first American whaling ships rounded Cape Horn and entered the Pacific

Frame 2: Whaling and the Opening of Japan

Japan was still a closed country when foreign whalers appeared in her coastal waters. Japan had kept open a window towards the western world only through Nagasaki where the Dutch were allowed to conduct a very restricted form of trade (Toby 1984), but Japanese were not allowed to leave the country and foreigners were not allowed to set foot on Japanese soil. This situation soon led to conflict between Japan and a number of foreign powers, and the presence of western ships off Japan's coast alerted the feudal authorities, prompting them to build coastal batteries and to increase the corvée labour laid on the fishing villages in connection with the establishment of a coast guard (Kalland 1988).

These precautions were, however, belated. The western powers had been enraged over the harsh treatment received by some wrecked whalers stranded in Japan. Moreover, they wanted access to Japanese harbours in order to obtain supplies of coal, water and fresh food for their whalers. On several occasions foreign ships sailed into Japanese harbours in the hope of obtaining concessions from the Japanese, and bringing with them Japanese fishermen and sailors who had been rescued in the hope of softening Japanese attitudes. But they were singularly unsuccessful until 1853, when Commodore Matthew C. Perry sailed into Tsuruga Bay with a fleet of four warships and issued the Japanese with an ultimatum. With this "battleship diplomacy" the Japanese were given one year to consider whether they wanted to have their country opened by peaceful means or by force. Japan gave in to Perry when the ships appeared the following year, and within a few years Hakodate and Yokohama were opened to foreigners.

The needs of western whalers were thus not only partly responsible for the Japanese government's decision to open the country's doors, but also led, a few years later, to the downfall of the Tokugawa regime.

in 1791.

These were sailing ships, typically between 300 and 400 tons, manned by crews of between 30 and 40 persons. The ships carried several small rowing boats which were lowered and used in the actual hunt when whales were spotted. These boats carried a crew of six: a helmsman in the stern, four oarsmen and the harpooner in the bow. One or two harpoons tied to long ropes were thrust against the whale at close range. Usually the whale was not killed by these harpoons but raced off with one or more whaling boats in tow. When exhausted, it was finally killed by the helmsman who was also a skillful handler of the lance which he threw into one of the whale's vital organs. The whale was then brought to the side of the ship where it was flensed. The blubber was stripped off and hoisted on board by pulleys. The heads of sperm whales were also hoisted on board, while the remains of the carcass, meat included, were dumped in the sea.

Unlike the British and Dutch who stored the blubber on board for later processing on land, the Americans returned to the old Biscayan method, long abandoned, of boiling the blubber in large brick furnaces on the whaling ship itself. This improved the quality of the oil, and enabled the whalers to remained at sea for an average of as long as three and a half years.

In a way, therefore, it is somewhat misleading to say that the Japanese tried to adopt the American way of whaling. After all, the American method and the existing Japanese method of net-whaling were too different to make wholesale adoption feasible. The Japanese operated from shore stations and made full use of the whales, and they were not prepared to change that. What they did do was try to make use of American hunting techniques, primarily by using bomb lances. The first experiments with fire-arms had been made by the British in the 1730s, but it took more than a century before they were used to any significant extent. From the 1820s a series of inventions culminated in the bomb lance in the late 1840s.

The Japanese had early on made contact with American and British whalers. Foreign whalers were seen in increasing numbers off Japan's coasts, and the sound of fire-arms could be heard from the shore. The first close observation might have occurred in 1823 when some Japanese allegedly boarded a foreign whaler off Hitachi. But also, in the next few years, a number of ship-wrecked Japanese sailors and fishermen were rescued by foreign whalers. One of these was Jirōkichi who, in 1838, was picked up by the American whaler *James Loper* after having drifted for six months in the Pacific. He returned to Japan in 1843 and gave a detailed description of American whaling (Fukumoto 1978:75-76). The most celebrated case is, however, that of Nakahama Manjirō who was rescued in 1841 by the three-masted American whaling ship, *John Howland*, under the command of William H. Whitfield from Fairhaven in Massachussetts (Frame 3).

Frame 3: Nakahama Manjirō, a National Hero

Manjirō was only a boy of 14 when he accompanied four others on a fishing trip off the southern shores of Tosa (Shikoku), shortly after the New Year of 1841. Heavy waves broke the boat's oarlock and the boat drifted aimlessly for a couple of weeks before reaching a small barren island where it was smashed by the surf. The five fishermen managed to reach land, and here they eked out a precarious living for five months. Finally, on 27 June, they were rescued by the *John Howland*'s crew who were looking for turtles on the island.

Manjirō recovered quickly from the ordeal and was then sent to the mast-top to act as look-out for whales. He so impressed William Whitfield with his enthusiasm and willingness to learn that the captain sent him to live with his aunt and friends in Fairhaven. Here Manjirō received proper schooling and thus became the first Japanese student in America. He excelled at college, where he also studied navigation, before being employed on the whaling ship *Franklin*, becoming her first mate in 1848. But Manjirō longed to go home to his family in Tosa, and toward the end of January 1851—under cover of night—he and two of his original companions from Tosa slipped ashore on Okinawa. After imprisonment and long interrogations in Naha, Kagoshima, Nagasaki and Tosa, he finally came home early in October 1852.

Manjirō had been thoroughly questioned about American whaling and shipping, and before long he was called on to teach at the clan school in Tosa. Among his students were Iwasaki Yatarō (the founder of Mitsubishi), as well as Sakamoto Ryōma and Gotō Shōjirō (two leading figures in the Meiji Restoration). In 1854 he was called to service by the Shogunate and worked as an interpreter when Commodore Perry came back to Japan that year. He translated a number of foreign books on navigation, astronomy and ship maintenance, and also assisted the authorities in drawing up plans for modern sailing ships and in training sailors for the navy (cf. Plummer 1984).

After having studied in Fairhaven and working as first mate on an American whaler, Manjirō finally returned home in 1852. Because of his great knowledge of the foreign world, he was summoned by the authorities to do various jobs for them. Once he was ordered by the Shogunate to serve as the captain on the *Ichiban-maru* which was sent to Ogasawara to whale. Although they returned with only two whales

(Fukumoto 1978:206), this was an important event since it was the first known Japanese attempt at pelagic whaling using the American method. Manjirō was also sent to Hakodate in Hokkaidō around 1862, to teach the local people the American way of whaling, and in 1866 the magistrate's office in Hakodate encouraged people to seek employment in the foreign fleet in order to receive training (Iwasaki 1987:16-17).

A number of attempts were made throughout Japan to establish whaling enterprises based on the American hunting method. One of the more successful companies was Gotō Hogei which was established in Arikawa in 1884. Between 1884 and 1888 this company caught 230 whales. But most of these companies were unsuccessful and thus shortlived, so that the American method of whaling did not make any real impact on the development of Japanese whaling.

The Norwegian Method

While the Japanese were trying to adopt the American method, Svend Foyn from Norway developed what became known in Japan as the "Norwegian method" of whaling, a method which in fact revolutionized whaling throughout the world. Originally a sealer, Foyn had acquired great knowledge of Arctic waters before turning to whaling in 1864 with his new ship, the *Spes & Fides*, about 90 feet long and the first steam whaling ship in the world. His most famous invention, however, was the grenade harpoon which he started to try out in 1864 and finally patented in 1870. Unlike the bomb lances, which were hand-held, Foyn's harpoon-gun was heavy and had to be mounted at the bow of the whaling ship. His method of hunting whales thus differed from the American one in two important ways: firstly a mounted harpoon-gun was used, and secondly the whales were shot from the steam-powered whaling ship itself and not from small rowing-boats. This enabled the whalers to pursue and catch large, fast-swimming baleen whales—such as the blue, fin and sei whales—whereas traditional whaling conducted from rowing-boats had only been able to catch slow-swimming mammals like right, grey, humpback and sperm whales.

Foyn also parted with the Americans in his way of processing whales. Since he caught mostly large baleen whales off the coast of Northern Norway, there was little need to process the whales on board ship. Instead, he established several landing stations where he aimed at an almost total use of the whale carcass, since "the whale is a gift from God". Oil was by far the most important product, and it was extracted from the blubber, meat and bones. Some meat of sei whales was also used for food: either fresh, dried, salted, canned, or used in sausages and hamburgers. Residuals from boiling oil were used for various products: the blubber was used for fuel, the bones were crushed into bone-flour and used as fertilizer, the meat was dried, ground and turned into guano and cattle feed. Glue was a by-product of all oil extraction. These processing stations became a marked feature of modern whaling.

Foyn worked the whaling grounds off the coast of the northernmost Norwegian province, Finnmark, close to the Russian border. The success of his endeavours did not go unnoticed by the Russians who by 1884 had ordered three whaling ships from Norway and introduced the "Norwegian method" to the Murman coast. It did not take long before word spread to the Russian Pacific Coast, and in 1889 a whaling boat built in Norway left Kristiania (present day Oslo) for Vladivostok. On board were Samuel Foyn—a distant relative of Svend—as captain and his second cousin Hjalmar Bull as helmsman. Both were also gunners. The two Norwegians were laid off after the first season in 1891, as it was assumed that the Russians had learnt what there was to learn. Soon after, however, the boat and crew were lost, so that the first Russian attempt at catching whales by the Norwegian method came to nought.

After two more failures, the "Pacific Whale Fishing Company of Count H.H. Kejzerling" was established by Kejzerling. Two boats, the *Nikolaj* and *Georgij*, which some years later became important in the development of modern whaling in Japan, were bought in Norway, and with Norwegian gunners they started operations from Hajdamak, east of Vladivostok. During the summer season these boats hunted in the Okhotsk Sea, bringing the whales to land for processing, while during winter they operated near Korea, bringing the carcasses to Nagasaki where they obtained better prices (Tōyō Hogei 1910; Tønnessen and Johnsen 1982:132). The company expanded and purchased transport ships to carry the whale meat to Japan, and in 1903 the first factory ship, the *Michail*, came into operation.

It was through Russian whaling that the Japanese were first introduced to the Norwegian method, for several of the Japanese pioneers had been in Russian service. But it hurt Japanese pride to leave the whales to their arch-enemies so that several attempts were made to introduce the Norwegian method to Japan. Entrepreneurs from the prefectures of Nagasaki and Yamaguchi, which were both centres of whaling during the days of the net-method, were particularly involved.

The first attempt was made in Nagasaki when a company was established in 1896 on the initiative of a carpenter, T. Takahashi, who had been in Russian service. With Takahashi as gunner on board a small wooden vessel, however, the company failed to catch anything, so it was reorganized and became known as Enyō Hogei. A larger wooden boat, the *Saikai-maru*, was the first modern catcher boat built in Japan, and a Norwegian gunner, E.E. Walby who had been working for the Russians, became the first of a series of Norwegian gunners hired by Japanese whaling companies (Tønnessen 1967:197). Tainoura in the township of Arikawa was made the company's base, but the boat caught only three whales and the company was dissolved in 1898.

In the same year, an Anglo-Russian Company was set up under Holme-Ringer Co., a British company in Nagasaki. Their catcher boat, the *Olga*, was built in Norway and both the gunner and skipper were Norwegian. At about the same time Arikawa Hogei was established, taking over the *Saikai-maru* and renaming it *Hatsutaka-maru*.

Both chartered out their boats in 1901 to the two companies that came to dominate Japanese whaling in the following years. The newly founded Nagasaki Hogei chartered the *Hatsutaka-maru* and Nihon Enyō Gyogyō the *Olga*. Nagasaki Hogei had been established by entrepreneurs who had gathered experience from previous whaling attempts in Nagasaki while Nihon Enyō Gyogyō was set up in Yamaguchi, where plans to start modern whaling were as old as in Nagasaki. However, for various reasons these plans did not materialize until Oka Jūrō, who came from an old whaling centre in Japan, was contacted by Fukuzawa Yūkichi, the founder and rector of Keiō University where he had studied. Fukuzawa wanted him to develop the rich marine resources off Korea and when Oka presented his plans for a modern whaling company to the Government in Tōkyō, the latter immediately endorsed them because it had, since the Sino-Japanese War in 1894-95, been trying to strengthen certain strategic industries—including pelagic fisheries which were stimulated as a means of gaining influence in the Japan Sea and the Korean Peninsula and so meeting what the Japanese saw as the Russian threat (Tōyō Hogei 1910).

Nihon Enyō Gyogyō was founded in 1899 and Oka Jūrō was soon after sent on a study trip to Norway. A new boat, the *Chōshū-maru No.1*, was built in Japan and with a Norwegian gunner caught its first whale in February 1900. After losses during the first season, there was a nice profit the following year. The future looked bright and for the 1901-02 season the company chartered the *Olga*. But *Chōshū-maru No.1* was wrecked in a blizzard and only Oka's stubbornness saved the company from being dissolved. With two boats chartered from Norway, the company had spectacular catches the following three years.

The Russo-Japanese War in 1904-05 had important impacts on Japanese whaling in several ways. Firstly, the Russians were expelled from Korean waters, so that Japan was more or less able to monopolize the whaling grounds between Taiwan and Ogasawara in the south and Sakhalin in the north. Secondly, most of the Russian whaling fleet was captured by the Japanese and the Japanese government then handed the boats over to Oka's company, which changed its name to Tōyō Gyogyō. With these news boats Tōyō Gyogyō was able to start catching whales in the waters off Tateyama and Chōshi in Chiba Prefecture and Ayukawa in Miyagi Prefecture in the summer of 1906, with great success. Nagasaki Hogei, which was desperately trying to compete with Tōyō Gyogyō, bought four new ships in 1906 and 1907, and as a result increased its catches many times over. Thirdly, with the elimination of the Russians and with increased catches by these two companies, a number of new companies were established during the years following the war. In 1908, twelve companies were operating 28 v haling boats from 22 shore stations in Japan and Korea and together they caught a total of 1,784 whales.

The number of boats increased further and the Japanese Government took its first steps towards regulating whaling, thus taking an active part in forming a "plan rational" economy,[4] which could both secure resources for the future and avoid over-capitalization. The companies were thus ordered to cooperate and merge into more viable units and in 1909 the four leading companies joined together to form

---

[4] For a discussion of plan rational *vis-á-vis* market rational economies, see Johnson (1982).

Tōyō Hogei, buying up two smaller companies in the process.[5] This new company controlled two thirds of the whaling fleet and 20 land stations. The Japanese Whaling Association (*Nihon Hogeigyō Suisan Kumiai*) had been established the previous year, its aims being to develop the whaling industry, conserve stocks and improve the industry's overall earnings,[6] and all companies were obliged to become members of the association. The following year the Government enacted the first whaling regulations in Japan. Licences were required and the number of licences was first limited to 30, a figure which was reduced to 25 in 1934. The new regulations also covered the species to be hunted, as well as whaling seasons and hunting areas (le Grand et al., n.d., p.4).

It was through such steps that modern coastal whaling for large-type whales (LTCW) can be said to have been firmly established in Japan. The industry was on a secure footing and expertise was gradually being built up. Norwegian influence had been strong, and Norwegian gunners continued to be employed by the Japanese companies for another 30 years, but Japanese LTCW had nevertheless developed its own distinctive characteristics. Whale meat for human consumption remained its important product, and this required different processing procedures from those found in Norwegian whaling. Oka Jūrō was well aware of this fact, when after his trip to Norway he said that Norwegian methods should be adopted as far as hunting was concerned, but that "when it comes to the processing of the meat we have kept the old system unchanged... (Japanese whaling) is a blend of European and Japanese elements and the result is a form peculiar to our country" (Tōyō Hogei 1910).

To say that LTCW was firmly established, however, does not mean that the industry did not undergo important changes in the following decades. These changes can be summarized as follows: firstly, new shore stations with flensing facilities were built to replace the flensing ships which initially had been used. Secondly, new whaling grounds were opened further to the north and round the Ogasawara (Bonin) Islands in the south. Whaling off the Kuril Islands

---

[5] The four companies were Tōyō Gyogyō, Nagasaki Hogei, Dainihon Hogei, and Teikoku Suisan and the two companies that were bought up were Tōkai Gyogyō and Taiheiyō Gyogyō (Tōyō Hogei 1910).

[6] These were, incidentally, the objectives put forward by the IWC when it was formed about 40 years later.

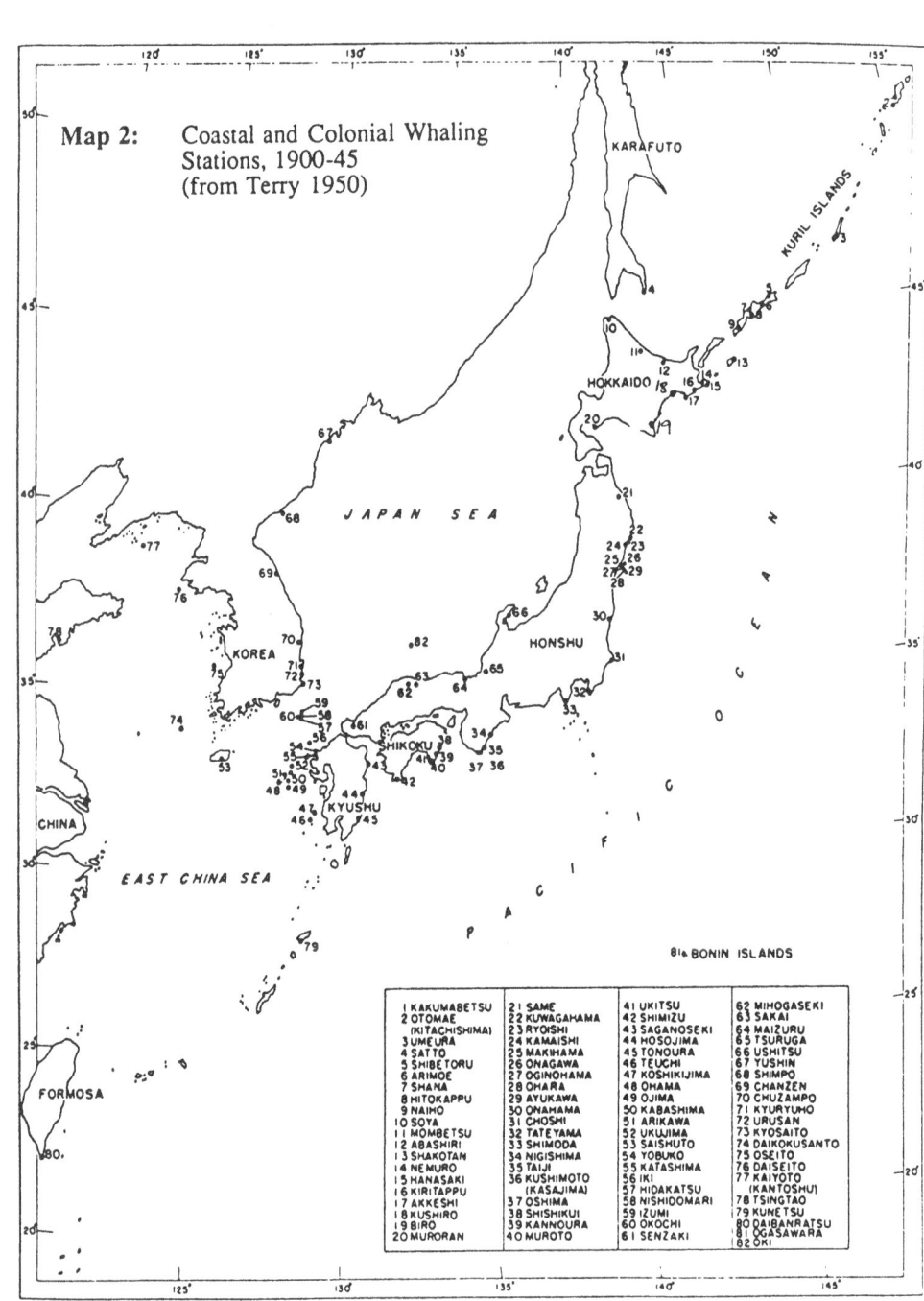

Map 2: Coastal and Colonial Whaling Stations, 1900-45 (from Terry 1950)

1 KAKUMABETSU	21 SAME	41 UKITSU	62 MIHOGASEKI
2 OTOMAE (KITACHISHIMA)	22 KUWAGAHAMA	42 SHIMIZU	63 SAKAI
3 UMEURA	23 RYOISHI	43 SAGANOSEKI	64 MAIZURU
4 SATTO	24 KAMAISHI	44 MOSOJIMA	65 TSURUGA
5 SHIBETORU	25 MAKIHAMA	45 TONOURA	66 USHITSU
6 ARIMOE	26 ONAGAWA	46 TEUCHI	67 YUSHIN
7 SHANA	27 OGINOHAMA	47 KOSHIKIJIMA	68 SHIMPO
8 MITOKAPPU	28 OHARA	48 OHAMA	69 CHANZEN
9 NAIMO	29 AYUKAWA	49 OJIMA	70 CHUZAMPO
10 SOYA	30 ONAHAMA	50 KABASHIMA	71 KYURYUHO
11 MOMBETSU	31 CHOSHI	51 ARIKAWA	72 URUSAN
12 ABASHIRI	32 TATEYAMA	52 UKUJIMA	73 KYOSAITO
13 SHAKOTAN	33 SHIMODA	53 SAISHUTO	74 DAIKOKUSANTO
14 NEMURO	34 NIGISHIMA	54 YOBUKO	75 OSEITO
15 HANASAKI	35 TAIJI	55 KATASHIMA	76 DAISEITO
16 KIRITAPPU	36 KUSHIMOTO (KASAJIMA)	56 IKI	77 KAIYOTO (KANTOSHU)
17 AKKESHI	37 OSHIMA	57 HIDAKATSU	78 TSINGTAO
18 KUSHIRO	38 SHISHIKUI	58 NISHIDOMARI	79 KUNETSU
19 BIRO	39 KANNOURA	59 IZUMI	80 DAIBANRATSU
20 MURORAN	40 MUROTO	60 OKOCHI	81 OGASAWARA
		61 SENZAKI	82 OKI

commenced in 1913; in 1919 whaling started off Taiwan and in 1922 at Ogasawara. In 1935, 38.6 per cent of the large-type whales were landed in Hokkaidō or the Kuril Islands, 39.3 per cent were landed in Tōhoku (Ayukawa and Kamaishi), 4 per cent at Ogasawara and only 1.4 per cent in Kyūshū (Appendix 1).

Thirdly, with whaling stations spread over an area between Ogasawara at 25°N and the Kuril Islands at 50°N (Map 2), the Japanese whaling boats could operate all year. The winter seasons were spent in the south, around Ogasawara, Taiwan and southwestern Japan, and the summers were spent in the north.

Fourthly, although the number of large-type whales caught in Japanese waters remained fairly constant between 1911 and 1945, with an average of 1,661 whales annually, the relative importance of the various species taken changed over time. The numbers of blue and fin whales decreased sharply while the number of sperm whales more than offset this decline. At the same time, this shift in species influenced the distribution and consumption of the whale carcasses to some degree.

Finally, the number of whaling companies contracted further. Several of the remaining companies were absorbed by Tōyō Hogei, which in 1934 merged with Nihon Sangyō to form Nihon Hogei. Two years later, this company merged with the fishing company Kyōdō Gyogyō, to form a new company known as Nihon Suisan (Nissui). Three smaller companies, based in the old whaling district of Tosa in Shikoku, were absorbed by the large fishing company Hayashikane Shōten, which established a subsidiary whaling company, Taiyō Hogei, in 1937. Although a few new companies were founded, they were all rather shortlived. One of these was Ayukawa Hogei, established in 1925 but bought by Kyokuyō Hogei in 1937. By the end of the 1930s LTCW was firmly controlled by three large companies: Nissui, Taiyō and Kyokuyō—all three being established, in fact, to finance pelagic whaling in the Antarctic. (For further details, see Chapter 8.)

Pelagic Whaling

Pelagic whaling had been conducted in Europe since medieval times when the Basques dominated whaling. The Dutch, British, and finally the Americans then brought pelagic whaling to most of the great oceans, but it was the Norwegians who opened up the last whale sanctuaries in the Antarctic with their modern fleet of factory ships and catcher boats.

The first Norwegian factory ship was used in northern Norway in 1881. In 1890 and 1891 Foyn sent one factory ship with a catcher boat to Spitsbergen and the following two years he sent another to Iceland. The greatest incentive to develop floating factories came, however, with the rapidly declining catches of large whales in Norwegian coastal waters, as a result of which the whaling grounds moved further north. A makeshift factory ship was brought to Spitsbergen in 1903. Two years later eight companies operated factory ships at Spitsbergen as a result of a ban on LTCW introduced in Norway that year (Tønnessen and Johnsen 1982:99).[7] The same year the first factory ship, the *Admiralen*, was sent to the Antarctic.

Several Norwegian expeditions had been sent to the Antarctic from 1892, partly to look for whales, and partly to catch seals. These expeditions did not catch many whales but they did pave the way for the first successful whaling in the Antarctic. In 1904, C.A. Larsen from Norway initiated Antarctic whaling when he founded a whaling station on South Georgia for the Compania Argentina de Pesca S.A. The following year Chr. Christiansen sent the *Admiralen* to the Antarctic. Then, in 1924, the Norwegians developed a slipway in the stern of the factory ships, allowing the whale carcasses to be winched up to the flensing deck. This enabled operations to be completely independent of land facilities, and pelagic whaling in the Antarctic was firmly established.

Considering the rapid modernization of Japanese LTCW after 1906, it might come as a surprise to find that the Japanese did not enter pelagic whaling earlier. In 1929, Tōyō Hogei bought a ship in England with the intention of sending it to the North Pacific after first

---

[7] The whale resources off Northern Norway declined and the whaling boats had to go further and further out to sea to catch whales. At the same time, the fisheries on which the local fishermen depended for their living failed. The fishermen believed that whaling was the cause of this, arguing that with fewer whales in the ocean, fewer fish would be driven toward the coast. Their antagonism against whaling was not helped in that the whalers came from Vestfold in southern Norway, the province from which all the Norwegian fleets to the Antarctic were sent. In 1903 some thousand angry fishermen attacked and levelled one of the land stations in the north. The following year the Norwegian government gave in to the fishermen's claims. (Cf. Olsen 1982:26-30; Tønnessen and Johnsen 1982:61-67.)

refitting it as a factory ship. The collapse of the oil market in 1930-31, however, forced the company to give up its plans and the ship was scrapped in 1933.[8]

The crisis in the oil market also forced the Norwegian whaling companies to withdraw their fleets from the Antarctic in 1932. One of these factory ships, the *Antarctic* (built in 1906), was offered for sale with its five catcher boats to the newly formed Japanese company Nihon Hogei. The ship was renamed the *Tōnan-maru* and was sent by the company to the Antarctic for the first time in 1934. Between seven and twelve Norwegians were hired as experts. The fleet returned to Japan after less than two months in the Antarctic and after catching only 213 whales.

This first fleet was faced with a number of difficulties. The purchase of the ships had met with serious opposition in Norway and other western countries which preferred to have Japan excluded from the rich catching grounds in the Antarctic.[9] At home in Japan, Nihon Hogei had difficulty in hiring crews since many whalers feared that they would never see Japan again. Moreover, at that time many whalers had to get permission from their fathers if they wished to work in the Antarctic. Nevertheless, even though several of the whalers died of *beriberi* during the expedition, the *Tōnan-maru* was sent south again the following year.

Japanese Antarctic whaling expanded rapidly in the following few years. Nihon Hogei/Nihon Suisan received its sisterships, the *Tōnan-maru No.2* and *Tōnan-maru No.3* in 1937 and 1938 respectively,

---

[8] Pelagic whaling in the Antarctic had expanded rapidly from 1904 when the first Norwegian fleet was sent. The invention in 1924 of the slipways at the rear of the mother ships, was another boom to this kind of whaling. In 1930-31, 41 mother ships and 6 land stations operated in the Antarctic with altogether 232 catcher boats. About 31,500 blue whales were killed in that season and more than 3.5 million barrels of oil were produced (Tønnessen and Johnsen 1982:386-87). These were the highest figures ever recorded for a single season, and this, together with the international depression, caused a slump in oil prices. Restrictions on whaling were introduced and the number of Norwegian factory ships in the Antarctic was reduced from 27 in 1930-31 to only 11 in 1933-34.

[9] When the Norwegians reduced the number of factory ships after the collapse of the whale oil market in 1931, Japan was offered one of the old ships. Some Norwegian newspapers and politicians, however, argued against the sale because it was thought that this would bring another competitor into an already saturated market.

while Taiyō Hogei got its first factory ship, the *Nisshin-maru*, in 1936 and the *Nisshin-maru No.2* the following year. Kyokuyō Hogei sent its new *Kyokuyō-maru* to the Antarctic in 1938, so that during the 1938-39 season Japan sent six fleets consisting of a total of 49 catcher boats and 8 transport and refrigerator ships. All the new factory ships were built in Japan, with the *Nisshin-maru* being completed in just four months.

In the summer of 1940, the *Tōnan-maru* and four catcher boats made up the first Japanese fleet sent to work the whaling grounds of the North Pacific, and the same fleet was sent again the following year so that, by the outbreak of the Pacific War in 1941, Japan had emerged as one of the world's leading pelagic whaling nations.

Small Type Coastal Whaling (STCW)

LTCW and pelagic whaling attracted much attention from the general public. The scale of operations was large, requiring huge capital investment and a large work force, as well as complicated processing facilities and administrative structures. Men's daring fights with the earth's largest mammals were also viewed with awe and fascination. STCW, on the other hand, has never been as spectacular as LTCW or pelagic whaling and as a result has received scant notice. The history of STCW, too, is not well known, even though it has been carried out for at least as long as LTCW. Through much of its history STCW has been conducted in two main ways: by nets and by the use of harpoons.

Japanese fishermen have for centuries driven schools of small toothed whales—such as dolphins, porpoises, pilot whales and killer whales—into bays where they were trapped. Such catches were seldom the result of long preparations, however. Rather, fishermen collected what fishing nets they had in an *ad hoc* fashion whenever a school was spotted, jumped into their boats and surrounded the cetaceans on three sides. They lowered their nets outside the school, and with loud shouts and beating on the sides of their fishing boats, herded the whales toward sandy beaches where they could be easily beached. Although dolphins and pilot whales are still occasionally caught in nets, the development of the harpoon method is more important for an understanding of STCW as it is found in Japan today.

It is likely that many of the harpoons which have been found in the shell mounds dating from the Jōmon period (10,000-300 B.C.) were used primarily to catch small cetaceans. Skulls of porpoises have been found arranged in such a way that one is led to believe that they may well have been used in religious ceremonies (Iwasaki 1987:9).

Frame 4: Catching Dolphins With Nets

A merchant, who resided by Hakata Bay (Fukuoka Prefecture) recorded the following events in his *Kenbun Ryakki* ("Notes on things seen and heard"):

"On the morning of the 14th day of the 8th month (1863), a school of several hundred dolphins was spotted from Karatomari Village in Hakata Bay. A number of fishing boats were sent out from the village, and gradually they drove the dolphins toward land. The boats carried a number of fishing nets used to catch high-quality fish such as yellowtail and Spanish mackerel, and these were cast one outside the other. Coming close to shore the fishermen cast a net made only of ropes inside the other nets, and people on shore pulled this net towards them as though it were a beach seine. The dolphins rushed onto the beach where large crowds of people waited. Between five and eight persons were needed to land one dolphin. By nightfall the nets had been cast five times and altogether 94 dolphins had been caught, ranging in size from 60 to 1,200 kgs." They were classified as common porpoise (*nezumi iruka*) and *nyūdō iruka* (unidentified porpoises).

"Ten of the animals were given to the authorities and villages nearby. Another 30 were sold locally, while 49 were sold at auction. [This leaves five animals unaccounted for.] These animals were later re-sold to Imari near Karatsu (in Saga Prefecture). Of the sale, half of the money was given to the net-owners, who had several of their nets destroyed during the hunt, and of this part the boat-owners received 30 per cent. The other half was distributed throughout the village."

There were other ways to distribute the earnings from dolphins. When six dolphins were caught in Hamasaki, also by Hakata Bay, in 1867, one dolphin was shared by all the villagers, and the other five handed over to the young men's association (*wakamono*) for sale. This association gave some of the money earned to the group of middle aged men (*chūnen-chū*). The young men's association spent its share on dance and entertainment, whereas the middle aged donated their share to the local Kumano Shrine.

Some communities continued until recent times to specialize in using harpoons to catch small cetaceans such as dolphins, pilot whales and Baird's beaked whales. Taiji and Katsuyama were such villages, and it is to these two villages that we have to turn in order to find the roots of modern STCW. Taiji and Katsuyama whalers had taken pilot and Baird's beaked whales, respectively, for centuries, during which period the inhabitants had developed a liking for meats of these species. With the modernization of LTCW it was only a matter of time before the new technologies of powered boats and harpoon guns were introduced into STCW. The outcome of this process was the development of a new kind of boat and harpoon, which proved efficient for catching not just pilot and Baird's beaked whales, but minke whales as well.

In pre-modern times, the net whaling operators along the Kumano coast (including Taiji) often allowed their whalers to catch small cetaceans such as pilot whales and dolphins outside the net whaling season. This hunting of small cetaceans was different from net whaling operations in that it involved individuals, rather than organized groups, who worked independently as and when they felt like it, and who used hand harpoons from individually owned boats crewed by small groups (generally no more than seven men, a considerably smaller crew than that found on a net whaling boat). After the collapse of net whaling in Taiji following the disaster in 1878, this traditional form of pilot whaling became of greater importance there. The introduction of the semi-diesel engine and the invention of the five-barrelled harpoon gun in 1904 by Maeda Kenzō (who had, like so many other Taiji men, left for America where he learnt to be a gunsmith), led to traditional vessels (in Taiji called *tentōsen*) being outfitted with these innovations, and pilot whaling became a viable form of whaling, continuing into the 1970s.

The whalers in Katsuyama never adopted the net method, but continued to hunt Baird's beaked whales by harpoons throughout the Tokugawa period. A group of fishermen was first organized for whaling activities by the Daigo family in 1612. Daigo supplied ropes and food during the whaling season, and in return obtained the blubber from any whale caught. The fishermen, for their part, used their own boats and harpoons, and shared the meat with the flensers and other local people who helped with the processing of the whales. Local women and children were allowed to scrape the bones free of meat. The Daigo family received from the feudal authorities exclusive rights to hunt Baird's beaked whales, and whaling was conducted by this group without interruption until 1870 when declining catches forced the group to suspend its operations temporarily. Whaling continued thereafter, on and off, until about 1905 when it came to an end in Katsuyama. Baird's beaked whaling, however, was taken up a few years later by Tōkai Hogei from nearby Tateyama City, using the Norwegian method. The company employed a number of Katsuyama fishermen and the meat found a ready market since local people were used to this kind of food. Tōkai Hogei continued its activities from various stations along the Bōsō Peninsula until 1969.[10]

By the early 1910s, two viable forms of STCW using modern harpoon guns had emerged in Taiji and on the Bōsō Peninsula. But the catcher boat for hunting minke whales had yet to be developed. This kind of boat, which eventually came to replace the boats used for pilot and Baird's beaked whales, was developed in Ayukawa along the lines of the Taiji *tentōsen*.

In the early 1930s, a seven ton pilot whaling vessel was brought to Ayukawa from Taiji and, by fitting this boat with a newly introduced 26mm Norwegian harpoon gun, it became possible for the first time to hunt minke using a small vessel. The Norwegian gun was mounted behind the Maeda gun which was used to fire the first harpoon; the 26mm gun was used to fire the second, and fatal, shot. Experimentation in Ayukawa led to modifications in the boat's design, and eventually the Maeda and 26mm guns were replaced with a more powerful 50mm harpoon gun of Norwegian design. This new design proved its worth and led to the general adoption of small-scale whaling boats (generally 15 to 20 tons) to catch minke and Baird's beaked whales (Ōmori, ms.).

Japanese Whaling in the Post-War Period

With the outbreak of the Pacific War in 1941, Japanese whaling faced grave problems. It became impossible to continue pelagic whaling, since all the factory ships and most of the catcher boats were taken over by the authorities and used in the war effort. Nevertheless, whaling was regarded by the authorities as an important activity in

---

[10] The above paragraph is based on Appendix 2 (written by Takahashi Junichi) in Akimichi et al. (1988:86).

order to feed a population increasingly faced with severe food shortages, and special considerations were given to the coastal whaling fleet. Thus, the productivity of LTCW in fact reached a peak in 1944 with production rising to 34,800 tons, of which 28,600 tons were used as meat. As an emergency measure the STCW boats were also allowed to hunt large-type whales and this form of whaling consequently expanded during the war.

By the end of the war Japan had lost 94.6 per cent of her whaling vessel tonnage, as all her factory ships and most of the catcher boats had been sunk. Moreover, more than half the prewar whaling grounds were relinquished when the Japanese army had to retreat from Korea, Taiwan and the Kuril Islands (Maeda and Teraoka 1952:33-35). At the same time, the population as a whole became more dependent than ever on whale meat because of the food crisis in the immediate postwar years. The occupation forces (SCAP) therefore decided to give special consideration to Japanese whaling, and General MacArthur authorized Japan's re-entry into international whaling (Hoel 1986:64). As a result the industry recovered remarkable quickly.

STCW was not adversely affected by the war defeat. On the contrary, the expansion of the STCW fleet continued during the early postwar years and reached a total of 83 boats in 1947. The Japanese government therefore began to rationalize this type of whaling through regulations at the end of that year. STCW was legally defined according to the species caught, the size of boat used and the distance from land in which it operated, and whaling licences were issued. In these new whaling regulations the size of the whaling boats was limited to 30 tons (increased to 47.99 in 1964), and catches of large whales by these boats were forbidden. In all, 86 small whaling boats were issued licences.

LTCW had not been that fortunate during the war, but managed to recover in a couple of years. The production of LTCW had already reached its prewar level by 1947, but remained thereafter fairly constant until the early 1970s (see Appendix 2), although the number of catcher boats in the same period declined from 42 in 1952 to only 12 in 1972 (le Grand et al., n.d.). During the 1950s, close to 20 shore stations were in operation. Most of these were located in the northeastern parts of the country after the Washington Convention of 1947 limited whaling to a six month period (i.e. between 1 May and 31 October). Since most whales appeared in the the southern waters during the winter, this decision meant the end to most whaling in the southwest of Japan.[11]

Pelagic whaling was by far the most important form of whaling during the late 1930s, with Antarctic whaling producing more than all other kinds of whaling put together. It was important both for the Japanese authorities and for SCAP to help this industry back on its feet, so the latter gave Japan permission to resume Antarctic whaling. In 1946 Nissui and Taiyō were able to send one fleet each with remodeled oil-tankers serving as factory ships (Japan Whaling Association 1954:22-24). These were replaced by new factory ships, the *Tōnan-maru* and *Nisshin-maru*, in 1951. By 1960-61, Japan was sending seven fleets to the Antarctic and, in the following season, production reached an all-time peak with more than 300,000 tons of oil and meat.

The recovery of pelagic whaling in the North Pacific was somewhat slower, although Japan was allowed to resume whaling operations off Ogasawara (Bonin Islands) as early as the 1945-46 season. This sei whaling ground had before the war been exploited from shore stations, but the Occupation authorities did not allow the Japanese to re-open these. Nissui was therefore unable to start whaling off Ogasawara that year, although Taiyō managed to take up whaling on 1 March 1946, using an old naval vessel as a factory ship (Shimada 1947). Two years later both Nissui and Kyokuyō had joined Taiyō, but since it was much more expensive to operate with factory ships than from shore stations, the whaling operations were not profitable. After 1950, therefore, only Kyokuyō remained in these whaling grounds. In 1952 whaling off the Ogasawara Islands ceased for a while when the IWC defined this whaling as coastal whaling, and so applied the same seasonal restriction as that to LTCW in Japan, with catches of whales being prohibited between November 1st and April 30th every year—exactly the season when sei whales appear around these islands.

However, 1952 was also the year in which the first post-war pelagic fleet was sent to the North Pacific. Two years later, two fleets were sent and from 1962 three fleets. Production reached its peak in

---

[11] A few stations were opened for shorter periods. Nissui tried to reactivate the land station in Senzaki (Yamaguchi Prefecture) in 1952, but gave up after two years. Taiyō was more successful when it opened a station at Arakawa in the Gotō Archipelago in 1955 for whales caught in the East China Sea. The catches of fin whales were spectacular, and Nissui opened another station in the archipelago (Tomie) the following year (Tønnessen and Johnsen 1982:661-662).

1967 with more than 90,000 tons.

In 1955 production surpassed the pre-war peak of 150,000 tons (1941) for the first time. But the relative importance of the various whale products had changed dramatically. Whereas 23 per cent of the production in 1941 was meat, this figure had risen to 44 per cent in 1955. Indeed, the production of whale meat in 1947 surpassed the pre-war peak in 1941 by 30 per cent, but that of oil did not surpass the 1941 level until 1957 (le Grand et al., n.d., p.186). The emphasis on meat production after the war meant that about 47 per cent of the total animal protein consumed by the Japanese in 1947 came from whale meat, and this percentage was still as high as 23 per cent in 1964.

Japanese whale production continued to increase, thanks to purchases of foreign fleets and quotas. By 1960 Japan had surpassed Norway as the leading whaling nation, but it was obvious to all that resources were being seriously depleted and that regulations were necessary. Japan had since 1909 regulated LTCW activities at home, and from 1947, all types of Japanese whaling required licences which were not easy to obtain. In 1946, too, the International Convention of the Regulation of Whaling was signed in Washington, with the IWC being established two years later. Although initially barred from membership of the IWC, Japan became a member in 1951 at a time when she was regaining her independence from the United States. These regulations were, however, inadequate to protect the resources from depletion.

As noted in Chapter 1, the interesting point about the IWC is that it was created not so much for purposes of conservation as for protecting the price of whale oil. This was reflected in the fact that quotas were not set for each species according to the conditions of the stocks, but in "blue whale units" (BWU). In terms of BWU, one blue whale equalled two fin whales, 2.5 humpback, 6 sei or 30 minke whales. Sperm whales were not counted in BWU, however, and catches of this species remained unrestricted until 1971. With no individual quotas set on the various fleets, it became essential to catch as much as possible before the total quota was taken. In other words, whalers found themselves involved in a zero-sum game. The more whales one fleet caught, the fewer were left to the others. What became known as the "Whale Olympics" had started. It became more rewarding for whalers to catch the largest species rather than waste time on the smaller ones, so that paradoxically, the regulations imposed by the IWC worked against, rather than for, conservation.

Finally, from the 1960s regulations became stricter. In 1961 the distribution of the total Antarctic quota was fixed at 33 per cent to Japan (an allocation Japan was able to increase by buying fleets—including quotas—from other whaling nations), 32 per cent to Norway, 20 per cent to the Soviet Union, 9 per cent to the United Kingdom, and 6 per cent to the Netherlands. In 1963, regulations came into force forbidding the catching of humpback whales, while the blue whale was completely protected from 1965. The total Antarctic quota was reduced from 15,000 BWU to 10,000 BWU in 1963-64, and further to 4,500 BWU in 1965-66, 3,200 BWU in 1967-68, and finally to 2,300 BWU in 1971-72. From 1972-73 the BWU system was abandoned and replaced by quotas for each species. Catches of sperm whales began to be regulated in 1971 and those of minke whales in 1973. From 1976-77, it was forbidden to catch fin whales in the Antarctic, and from 1978-79 the prohibition was extended to sei whales. This left sperm and minke whales as the only species hunted in the Antarctic.

In the North Pacific quotas for fin, sei, and sperm whales were introduced in 1969, and were thereafter steadily reduced. From 1976 hunting fin and sei whales was prohibited, and catches of minke whales were regulated from the following year.

Small quotas for certain species should have worked out well from a conservation point of view. But for reasons that we have discussed in the first chapter, in the 1960s and '70s whales became a symbol for many "animal rights" and environmental groups. At the United Nations Conference on Human Environment held in Stockholm in 1972, a resolution for a ten year moratorium on commercial whaling was adopted. Then in 1982, the IWC declared a total ban on all commercial whaling beginning with the 1985-86 season. Under the threat of an international boycott, the Japanese government decided to comply with the decision and sent its last fleet to the Antarctic in 1986. The shore stations serving LTCW were closed in 1987, and from 1988 the STCW boats were prohibited from catching minke whales. At present Japan only catches minke whales in the Antarctic for research purposes, and nine small coastal vessels are licensed to catch a few dozen Baird's beaked whale and pilot whales.

During the post-war years the structure of the Japanese whaling fleets has changed with the industry's changing fortunes. The three leading pre-war companies—Nissui, Taiyō and Kyokuyō—quickly re-established their hegemony after the war, despite the fact that Nissui was classified as a large financial clique (*zaibatsu*) and was thus forced to split up into several smaller companies. The company lost some licences to subsidiaries like Nittō Hogei (established in 1949) and Nihon Kinkai Hogei (established in 1950 and renamed Nihon Hogei in 1970). Several of the STCW boats owned and operated by Nissui were also transferred to subsidiaries and finally taken over by individual owners. The same process has lately been at work in companies like Taiyō and Nihon Hogei (cf. Chapter 8).

Until 1956, only Nissui and Taiyō were licensed by the Japanese authorities to catch whales in the Antarctic, but in that year Kyokuyō was finally allowed to participate. The Antarctic has been closed to other Japanese companies, although a few subsidiaries were invited to provide catcher boats for their parent companies. As for pelagic whaling in the North Pacific, the Japanese Government originally forced the three large companies to operate jointly until 1962 when they were finally allowed to send one fleet each, but with Nittō Hogei, Nihon Kinkai, and Hokuyō Hogei providing catcher boats and occasionally a factory ship to these fleets (cf. Chapter 8).

These six companies also dominated LTCW in Japan, but over-capitalization and low efficiency caused many of them losses. They therefore gradually reduced the number of boats from 42 in 1952 to only 12 in 1970, but without reducing catches as a result. From 1965 Kyokuyō pulled out of coastal whaling, while a subsidiary of Taiyō and Nihon Hogei, Sanyō Hogei, entered the business with one catcher boat in 1968.

STCW also saw a steady decline in the number of vessels being used. The small boats could hardly compete on the market with the larger and better equipped ships of the big companies when these had recovered their strength from the pre-war years. Thus, the number of STCW boats started to decline early: in 1951 there were 68 boats with licences; by 1954 their number had fallen to 54. At that time the Government decided to reduce the number further by converting a number of the licences to a single licence for a larger whaling ship of equivalent tonnage. 41 boats were scrapped between 1956 and 1961, bringing the total number of vessels down to 23. This reduction in the number of boats was also carried out without significant reduction in whale catches.

Until the early 1960s, Japanese pelagic whaling was expanding, while coastal whaling was rather stable. From the 1960s, however, the situation changed completely, as the industry contracted and so brought about a change in its structure. For a start, companies tried to open up new whaling grounds in distant waters. The three largest companies operated joint shore stations on South Georgia Island from 1963, since catches made by boats operating from land bases there were not included in the Antarctic quotas set by the IWC. When the catches from the shore stations on South Georgia were included in the quotas from 1967, the stations were closed down. In Canada, Taiyō operated a shore station at Coal Harbour, Vancouver Island, from 1962 to 1967, and Taiyō and Kyokuyō operated four stations in Newfoundland between 1966 and 1972, when they were closed by the Canadian Government. In Brazil, the Japanese companies Hokuyō Suisan and Nihon Reizoku entered into a joint venture with a Brazilian company in 1959 and processed the whales at a shore station near Recife. Hokuyō Suisan dropped out after the 1961 season, although Nihon Reizoku continued until 1976. Taiyō Gyogyō formed a joint venture with another Brazilian company and whaled from Imbituba further south between 1959 and 1963. Nittō Hogei did the same in Chile from 1963 to 1968 together with a Chilean company, and Nihon Kinkai conducted whaling from a shore station in Peru between 1967 and 1978 (Tatō 1985:48-58).

Such activities could not in the long run solve the grave problems facing the whaling companies. Cuts were necessary, and in the Antarctic the number of Japanese fleets declined from seven in 1964-65 to three or four between 1968 and 1975. The number of vessels engaged in coastal whaling had also decreased so that, by 1975, eleven boats were engaged in LTCW and only eight in STCW.

These reductions were, however, insufficient and the industry went through a major restructuring period during 1975-76. The whaling sections of the three pelagic companies—Nissui, Taiyō and Kyokuyō—were merged into a new company, Nihon Kyōdō Hogei. Minor shares were also held by Nittō Hogei, Nihon Hogei and Hokuyō Hogei. Kyōdō Hogei took over the three factory ships with 20 of the catcher boats, but had to scrap two of the factory ships in 1977 when the IWC decided to reduce quotas further. From that year Kyōdō Hogei operated one fleet both in the Antarctic and the North Pacific until the moratorium was enforced. The company was dissolved in November 1987, but in April the following year Nihon Kyōdō Senpaku was established in order to carry out research whaling in the Antarctic.

LTCW was also reorganized during these years. It was decided that Nissui and Taiyō should end their coastal whaling operations completely and they closed their last shore stations after the 1976 season. LTCW was left to the three remaining companies—Nihon Hogei, Nittō Hogei and Sanyō Hogei—but they were forced to close their shore stations in December 1987.

With the ban on minke whaling from 1988, STCW also underwent serious structural changes. The STCW operators formed partnerships, each consisting of two companies. One of the two boats owned by the partnership was withdrawn from whaling and only one remained in operation. Crews were fired, but a selected few were later re-hired to crew the boat in operation. A special situation emerged in Abashiri since the two boat-owners there let their partners in Ayukawa operate their vessels with their crews intact. This caused the lay-off of all the Abashiri whalers, except for two persons who were temporarily hired as flensers on the shore station there.

The number of whalers declined drastically as whaling operations contracted. This has had a serious social, economic and cultural impact on the whalers, their families, and the local communities in which they live. We shall deal with these impacts later on, but before we can fully appreciate all the implications of the declining industry and the present moratorium, we need first to analyse the social organization of whaling in Japan. This will be the subject of the following four chapters.

CHAPTER 5

Work Organization of Whaling

Let us now turn from the historical background of Japanese whaling and concentrate instead on the way work therein has been organized.[1] For it is by focussing on the work organization of whaling in Japan that we will be able to discuss the concept of a "whaling culture". Our basic argument is that any form of whaling can be broken down into a series of distinct components which represent stages of production. Here we will outline the main features of each of the three modern types of whaling defined in the previous chapter, our aim being to bring out structural similarities and historical continuities among different forms of whaling that have existed in the past and still exist today. In particular we will argue that there is a sharp division between activities involved in *hunting* and *processing* in all types of whaling, and that this division transcends differences among the three types of whaling under discussion. It is this which will enable us later to argue for the concept of a whaling culture in Japan.

Large Type Coastal Whaling (LTCW)

Large type coastal whaling is characterized by the species it pursues (sperm whales and the larger baleen whales [excluding minke]), by the scale of the boats used (often the same as those catcher boats used in pelagic whaling), by its reliance on land-based processing, and by the absence of factory ships.

---

[1] This chapter is to a large extent based on a paper that appeared in *Maritime Anthropological Studies* (Takahashi et al. 1989). We are grateful to the editors of the journal and to our co-authors for permission to make substantial use of that paper here.

In LTCW, each catcher boat (varying from 100 to just over 600 tons in size, with crews of roughly 20 persons) was a separate unit able and expected to make all decisions connected with hunting whales, including decisions about where and when to initiate a hunt. Within seasonal and geographical limitations imposed by Japanese government and other authorities, the gunner on the catcher boat decided which whaling grounds should be worked each trip, basing his decision on the extensive knowledge that he had accumulated over the years of seasonal migration patterns, as well as on information obtained from such natural phenomena as tides, currents and wind.

Once the boat reached the hunting ground, the actual search could begin. In general four men headed by the bosun gathered at the masthead to keep a look out for whales. Constantly monitoring the water temperature, as well as changes in water colour and wave patterns, the crew searched for whales where different currents met, knowing that that was where whales tended to satisfy their appetite on fish, krill, squid, and other creatures. The sighting of sea birds was also of great importance (as could be the presence of dolphins) since these signified the presence of whales in the area.

Ultimately, of course, the crew looked for the spout of a whale. An experienced whaler could tell from these spouts not only what species had been sighted and the direction in which the whales were moving, but, in some cases, how many were present. If the whale could be officially hunted (i.e. if it was permitted to hunt the species of that particular size in that area and season), the catcher boat would then embark upon the chase. Here traditionally there was very close cooperation between the bosun at the masthead and the gunner who stationed himself initially on the bridge of the catcher boat. The bosun sent instructions verbally (via a voice pipe or by microphone) to the ship's engineer to whom he relayed orders about engine speed and the direction of the vessel. It is important to note, however, that—even though he might delegate authority to the bosun in the early phase of the chase—it was in fact the gunner who was in charge of the catcher boat throughout the hunt. This became more obvious as the catcher boat closed in on its prey and the gunner moved forward to the harpoon platform to take over firm control of the final approach to the whale.[2]

In the past, both gunner and bosun needed to have as near perfect a knowledge of whale behaviour as possible for the pursuit to be successful. Later, however, their roles were modified by the invention of an echo sounder Asdic known as the *geitanki* (*lit.* "whale searching device"), introduced on all types of catcher boats from about 1960.[3] The Asdic is both a sonar-like device that can be used actively to locate whales through returned echoes, as well as an apparatus that can also passively receive the sounds of whales. The device was of use only after a whale had already been visually spotted and the boat had approached to within catching distance. Otherwise, if turned on too early, the signals emitted by the Asdic would scare the whale away. Once the whale was in range, catcher boats equipped with an Asdic could pinpoint precisely the presence of a whale, together with its direction and distance from the vessel. It was particularly useful when the whale dived and became invisible to the bosun, since—by tracing the path of the whale under water—it permitted the gunner to position his vessel perfectly for the final approach to the whale. Another advantage of this innovation was that it allowed catcher boats to follow whales throughout the night and so enable them to take up the final stages of the hunt the moment daylight returned.

More importantly, however, the invention of the Asdic affected the role of the gunner aboard the catcher boat during the chase, in that his detailed knowledge of whale behaviour and the likely movements of the type of whale being pursued were no longer as important as they used to be. This made the difference between good and bad gunners less obvious than it had previously been (although, as we have seen with the portrait of Tokunaga Yasuhiko, a gunner might not wish to admit this). Moreover, whereas in the old days it was the relation between gunner and bosun that was vital for a successful

---

[2] It appears that the role of the gunner in whaling is similar to that of the master fisherman (*gyorōchō*) in certain types of Japanese fishing, for the latter also takes over command of the vessel on the fishing grounds. For similar patterns of authority in fishing in other parts of the world, see f.ex. Barth (1966) and Zulaika (1981).

[3] This device was first invented by the allies during World War II in order to track down German submarines. Norwegian and British whaling companies quickly saw the potential in adapting the device for whaling, and it was used already in the first post-war season, after the military authorities had released this technology for civilian use (Tønnessen and Johnsen 1982:695-696).

pursuit, a new line of communication was now set up between the gunner, bosun and the sonar apparatus operator (*geitanshi* or *tsuigeishi*), who was himself a new addition to the composition of the catcher boat's crew.

All in all, the decline in the authority of the gunner on the catcher boats mentioned by informants appears to have coincided with the introduction and adoption of the Asdic. Indeed, the fact that considerable care was taken to ensure that the sonar operator did not infringe upon the sphere of influence hitherto wielded by both gunner and bosun indicates that this technological innovation brought about a potential source of conflict in crew organization aboard the catcher boats. For example, Asdic operators were careful not to make any statements that might suggest that they were issuing orders about the vessel's course, and instead restricted themselves simply to reporting the location of the whale. Moreover, the fact that the operator now gave information about the whale's movements over the boat's loudspeaker system meant that what was once *secret* knowledge—and hence one of the main sources of power supporting the gunner's authority—now became *shared* knowledge, thereby allowing other crew members to assess the performance of the gunner and of the bosun.

In order to shoot a whale, the catcher boat had to pursue it to within a range of 40 to 60 metres, after a chase of perhaps several hours before this close an approach became possible. In the final stages the gunner manoeuvred his vessel so that it approached the whale at an ideal angle of about thirty degrees to it. Depending on the species, the gunner may have had as little as two to three seconds in which to take aim and fire, but he also had to take into account such factors as the distance between the catcher boat and the whale, the absolute and relative speeds of the two, wind and wave conditions (preferred timing being when the bow of the boat was rising), and acquired knowledge of the behaviour of the whale itself (Tanaka 1987).

The next stage was marking and securing the dead whale. Large baleen whales such as blue, fin and sei were in general pumped full of air to keep them from sinking—something which was not necessary for whales rich in oil content such as right and sperm whales. The carcass was secured to a buoy and marked by a flag and, in later years, a radio transmitter and metal reflector for radar detection, before the catcher boat continued on its hunt for other whales.

Finally, the whale was brought back to land. It was the gunner's task to decide when to collect the whales and bring them back to the shore station, marking the end of a hunt which may have lasted several days or which may have been concluded in a single day if the hunting was successful. Here again, the species of whale had an important influence on his decision. Baleen whales, for example, had to be brought back to the shore station promptly since they were primarily consumed as fresh meat, and prices fell sharply with deteriorating quality; sperm whales, on the other hand, which were prized mainly for their oil, or used for preserved meat—either salted or canned—did not need to be towed to land so quickly. In some places like the East China Sea, where the water temperature is comparatively high and the carcass thus decomposed quickly, baleen whales were first bled by an incision in the neck and then had their entrails removed. In deciding when to convey the whales back to land, therefore, the gunner also had to take into consideration a number of factors, such as sea conditions, speeds of currents, the distance of the whales from the land, and of course the number of whales caught.

On being secured to the side of the catcher boat, whales of all species had the corners of their flukes cut, both to make their handling easier and to ensure that the carcass was not lost should wave movement happen to snap the tail off the carcass. All species of whales were also bled at this stage, if this had not already been done.

In LTCW the whale had to be processed on specially designated shore stations, and the station operators had to pay compensation to the local fishing associations for the inconvenience caused to fisheries by whaling operations (see Chapter 7). They also made frequent donations to the local community institutions as a good-will gesture. In this respect we can see that there was continuity between LTCW and pre-modern net whaling pactices.

Seven main processing activities were carried out at these land stations, although some of the tasks were sometimes subcontracted elsewhere: flensing, oil extraction, salting, icing of fresh meat, crushing of bones for fertilizer production, drying of sinews, and boiling of entrails for food. Subsidiary tasks occasionally undertaken by employees at shore stations included the cleaning of sperm whale teeth and baleen, for use in craft production, and the maintenance and repairing of flensing tools and facilities.

When the catcher boat reached harbour, it was met by a small tow-boat that came out to take the whale in tow as far as the slipway to the flensing station. There it was winched up on shore, tail first, the winches being operated by experienced workers who were not, however, exclusively specialists in this task alone. Flensers (*kaibō'in*) would often start their work while the whale was being dragged up by the winches onto the slipway, since they could thereby exploit the movement of the carcass in making the first cuts lengthways in it. Otherwise, they would wait until the carcass had been winched right up the slipway before making long cuts along its sides and, as soon as the carcass came to a rest, along the whole of the topside of the whale. They then cut off the tail, putting aside the fluke for slicing up later, before being salted (or transported to salting facilities if these were not available at the same station). The winches were next used to peel off the blubber from the head to the tail of the whale, while the flensers carefully separated the blubber from the meat. The blubber was put to one side, while the winch operators proceeded to peel the meat from the carcass as the flensers carefully separated the meat from the bone.[4] The meat was then cut into blocks 30 centimetres across, before it was further cut up into smaller chunks (sometimes by less skilled workers locally hired on a daily basis during especially busy periods) and placed in an ice tank for cooling. During this whole process, tendons and other tough parts were carefully removed to be used as scraps, and the whale meat itself was cooled with crushed ice.

Next, the blubber was similarly cut up into 30 centimetre wide strips. Since blubber had different uses, depending on the species of whale from which it was taken, some was used for salting and some for oil extraction. In the latter case, the large blocks were taken to the boiler section of the shore station, where they were further cut up and placed in the boilers to be prepared for oil.

Other portions, including the ventral grooves, dorsal fin, flukes, flippers (in the case of the humpback whale), skin of the whale, and—in the days before refrigeration enabled large quantities of red meat to be consumed as fresh meat—red meat in general, were all sliced and salted.

The remaining skeleton was sawn into pieces, before being taken from the land station to nearby fertilizer plants, which were often operated by local people, where it was crushed, dried and made into

---

[4] In the case of baleen whales, the *onomi* tail meat was removed first since it is considered a great delicacy in Japan (see below for further discussion of flensing techniques for different types of whale).

fertilizer. The sinews were also removed by a subcontractor who washed, stretched and dried them in preparation for musical instruments, tennis rackets, and so on.

Fresh intestines and other organs such as the heart, liver, oesophagus, and kidney were boiled, either at the shore station itself, or elsewhere by a subcontractor. If these entrails were not fresh, or if those employed on the shore station had no time to treat them, they were sent with the bones to be made into fertilizer.

Pelagic Whaling

In pelagic whaling, a similar procedure to that described for LTCW was followed, but there were certain important differences in the search and carcass collecting phases, on the one hand, and in processing, on the other. These differences were reflected in the composition of the pelagic whaling fleets, which varied somewhat from fleet to fleet, between catching grounds and over time. During the 1951-52 Antarctic season, for example, the (Taiyō operated) *Nisshin-maru* fleet consisted of a total of 23 vessels: the factory or "mother" ship (*bosen*), two salting/freezing ships, two freezing ships, four transport carriers, one tanker carrying diesel oil for the fleet, ten catcher boats, two towing boats, and one search vessel (Maeda and Teraoka 1952). In 1976, on the other hand, the Nihon Kyōdō Hogei fleet operating in the North Pacific, consisted of a factory ship and 9 catcher boats only. Both freezing and salting were done on the mother ship, and no other support vessels were needed because the hunting grounds were relatively close to Japan.

A major difference between LTCW and pelagic whaling, so far as hunting is concerned, was that the hunting phase of whaling was closely coordinated and directed by a commander-in-chief (*sendanchō*) from the factory ship. Basing his decisions on international whaling regulations and information on whale behaviour and sea conditions accumulated from previous years' whaling trips, the commander-in-chief first decided on the general area in which his fleet would pursue its whaling activities, and then sent out his search vessel to move ahead of the factory ship, reporting back by radio when sightings of whales were made.[5] Another search strategy, especially before the official hunting season for the economically more important baleen whales opened, was to hunt for sperm whales, using the hunting as an opportunity to carry out general reconnaissance. On the basis of information gathered during these various activities, the commander-in-chief deployed his catcher boats, ordering them to maintain a certain distance between themselves and to proceed towards the area in which whales had been sighted. From this point, however, the catcher boats took over the hunt and the search, pursuit and killing of the whales proceeded in exactly the same way as that described above for LTCW.

Another major difference between LTCW and pelagic whaling was to be found in the manner of securing and retrieving the carcass, once a whale had been killed. As in LTCW, the dead whale had to be secured with floats or pumped with air in order to prevent it from sinking, while the catcher boats proceeded with the hunting of other whales in the area. The carcass, however, was then collected and taken back to the floating factory by special towing boats. Also, before a catcher boat left the whale, its gunner made sure to attach a long bamboo pole to the carcass with a flag on top to identify fleet ownership and which catcher boat was responsible for the successful killing. A radio transmitter and metal reflector attached to the carcass enabled the collecting vessel to identify the whereabouts of the whale and tow it back to the factory ship for processing.[6]

A pelagic fleet operated on the open seas for months at a time. This influenced the range of products into which the whales were processed, as well as the work organization of the processing fleet, which included not only the factory ship for flensing and oil processing, but other ships where fine cutting, salting, and freezing were carried out. Such things as bones, which on land would have been processed into fertilizer or other products, were as much as

---

[5] This information was sent in code, so that each fleet spent a considerable amount of time and energy on trying to crack its rivals' codes. This meant that companies had to change their codes every second month or so. Such "negative information sharing" is important in fishing fleets in general, see Forman (1967); Andersen (1972); Stuster (1978); Byron (1980); and Gatewood (1984).

[6] When the scale of pelagic whaling contracted, the number of catcher boats was reduced and both search and collecting vessels were eliminated. This meant that the hunting operations in pelagic whaling became very similar to those found in LTCW.

possible, on board ship, processed into oil. Another slight difference was that at sea, of course, sub-contracting and the employment of extra workers during busy seasons were impossible. All in all, however, work on the fleet closely resembled that on the LTCW shore station.

The persons working on factory ships were organized into two main groups: the crew (*ōgata sen'in*) of roughly 90 who operated the ship; and an additional 250 managers and processing workers, who were subdivided between flensing and factory (mainly oil extraction) sections (Nihon Suisan 1966). There were two flensing decks on a mother ship, one at the stern where rough flensing was done, and the other in the centre of the ship where secondary cuts (*saikatsu*) were made.

As processing of a whale began, the carcass was winched tail first up the slipway at the stern of the boat by workers who were specially employed for this task. Then the flensers (*kaibō'in*) cut off the tail which was winched to the second flensing deck where it was cut up later by butchers (*saikatsu'in*) into smaller pieces for salting. The flensers, at least one on top and one on each side of the whale, cut the blubber along the length of the whale, before it was stripped off the meat by the winch operators.

The blubber was hauled by the winch to the front of the flensing deck where butchers cut it up in blocks 30 centimetres wide with the help of *kagihiki* ("pullers") who used hooks to spread the blubber as it was cut open by the butchers. Sometimes, the latter then separated the skin from the blubber so that the skin could be further cut up into 3 by 30 centimetre pieces for salting by other specialists (*enzō kakari*), either on the factory ship or on special salting ships. Most of the blubber was sent down through holes in the deck for processing in a Hartmann-type boiler by those employed below in the boiling section (*saiyubu*).

The next step in processing involved the flensing of whale meat. Since the *onomi* is particularly important in Japanese dietary tastes, the flensers were particularly careful when cutting this kind of meat, found near the whale's tail. They then separated the meat from the bones, an extremely skilled operation, after which the butchers proceeded to cut the meat up with the help of the pullers, who would ensure that the membrane covering the meat was always turned upside so that cutting was easier (again using hooks to help with the cutting). Until the late 1940s, the meat was then shipped by dories from the factory vessel to a separate vessel (*fuzokusen*) where it was further cut up into smaller pieces before being salted by about 180 workers employed there. In the late 1930s, however, freezing ships were introduced and this new way of handling the meat came to dominate from the late 1940s onwards, although freezing did not replace salting entirely. As whaling operations contracted in the late 1970s, floating factories were refitted so that, in addition to their previous functions, they were also able to freeze, salt and store meat, blubber, ventral grooves and entrails, until these were taken back to Japan by the transport ships that came to meet the Antarctic fleets.

As a final stage in the flensing operation, the remaining meat scraps were scraped off the skeleton by specialists, before the bones themselves were handed over to another set of workers who cut them into small pieces with chain saws. They were then crushed and, unlike in LTCW land stations where bones were often processed by special companies, put into a Kværner type boiler operated by people from the boiler section. The remaining entrails were also processed into oil.

It should be emphasized that the relative importance of the various products prepared by a pelagic fleet changed over the years as a consequence of changes in the market for whale products. In the 1960s, the demand for whale oil decreased. At the same time the number of captured whales also declined, which brought about higher prices for whale products processed for human consumption. This led to a marked shift in the use of the blubber from oil extraction to freezing and salting for food.

Small Type Coastal Whaling (STCW)

As we have seen, the hunting of small type cetaceans has been practiced in Japan in one form or another for many centuries, but the origins of what is now commonly referred to as small type coastal whaling (STCW) can be found in the beginning of minke whaling off the Japan coasts in the 1930s. This type of whaling is characterized, firstly, by the species caught (minke, bottlenose, beaked, pilot or killer whales), and secondly, by the use of powered vessels (in Japan maximum 48 tons) with mounted harpoon guns (IWC's schedule).

Although the hunting season is now fixed by Japanese government regulations, STCW has in fact always been practiced when whales are close to the coast. This means that the administratively regulated season has also been an ecological season. At the same time, the fact that the vessels are small means that STCW has been essentially a single day hunting operation, for the boat leaves its harbour in the morning on a clear day when the sea is calm and returns in the evening after dark. Only very rarely, when the sea is very calm, does it stay out overnight.

The crew of each whaling vessel is small compared with other types of whaling catcher boats, consisting of between five and eight persons (as opposed to a 16 to 23 member crew on a pelagic fleet catcher boat). Here the gunner has had even greater influence than was the case in the LTCW and pelagic operations, and may often be gunner, captain and owner of the boat all at once. Other crew members consist of an engineer and deckhands only. No specialized communications officer is employed, even though there is, of course, advanced radio communications equipment on board the whaling vessel. It will be appreciated that this lack of specialization among other members of the crew contributes greatly to the overall authority of the gunner who takes over complete control of the vessel once it leaves port.

As in LTCW, the gunner first decides the general area in which he will conduct his daily search. This will be based on past experience, seasonal variations of currents, whale migrations and general availability of food for the whale, together with information supplied daily by local fishermen. As in LTCW and pelagic whaling, he pursues his search carefully monitoring water temperature and the flow of the currents, while looking for other clues which also indicate the possible presence of whales—like the activities of birds, dolphins, and large fish. At the same time, he has to have a more specialized knowledge of the topography of the seabed than is the case in LTCW or pelagic whaling, since the STCW boats operate closer to shore and in shallower waters—the depth of which also affects the behaviour of the whales sought, especially the Baird's beaked whales. A whale will either be sighted directly or tracked on the basis of information received from fishing vessels at sea which relay news of sightings directly by radio to the whaling vessel (with the expectation of reward of whale meat should the information lead to a kill).

In STCW, both the structure of the whaling vessels and the type of whales hunted affect the way in which the chase is carried out. For example, a slow vessel chasing a minke whale may launch a small power boat which it sends out to slow down the whale and drive it eventually towards the whaling vessel.[7] A fast boat, on the other hand, obliges the whale to swim very quickly and so prevents it from diving—in which case it is overtaken by the whaling vessel on its own. Unlike minke whales, Baird's beaked whales dive as deep as 1,200 metres and for up to 45 minutes at a time. This means that the gunner has to try to work out where the whale will resurface and position his vessel accordingly. It is important to note that the Asdic used in LTCW and pelagic whaling is not used in STCW as a tracking device, in part because the beaked whale is extremely sensitive to its signals, and hence easily scared by it.[8] Therefore, the gunner's traditional—and secret—knowledge has remained extremely important to success in STCW during the pursuit of the whale. Moreover, his skills are tested much more fully when it comes to shooting a whale, since the target is smaller and the whaling vessel itself is much less stable (because of its small size) than were LTCW and pelagic catcher boats. Another peculiarity regarding STCW until recently was the use of the cold harpoon. Unlike the harpoons used for large type whales, it did not explode when it penetrated the whale blubber. Before the invention of a new exploding harpoon in the mid 80s, the whale was not killed instantly, and a second harpoon (*niban mori*)—sometimes aimed by the apprentice gunner—had to be fired. The introduction of the explosive harpoon made a second harpoon unnecessary in most cases, and in this respect modified the on-board training of new gunners to some degree.[9]

Once a whale has been caught the whaling vessel will usually secure the carcass by tying its tail to the side of the boat, bleed it, and then tow it back to the landing station. There are two exceptions to this rule. Firstly, if there are other whales nearby which the whaling vessel wishes to pursue, it will attach a radio buoy to the carcass,

---

[7] One Abashiri whaler claimed that with the use of this dinghy he was able to catch 90 per cent of the minke whales sighted, whereas only 60 per cent of them were caught without it. This innovation in fact resembles in some important respects the way in which whales were driven towards nets in pre-modern whaling by special fast boats known as *seko-bune*.

[8] STCW vessels do make use of a transmitting device which frightens minke whales, making them surface and swim fast. This device does not, however, have a receiver of the kind employed in LTCW and pelagic whaling catcher boats.

[9] The explosive harpoon was developed to meet international criticism that whales were being killed inhumanely.

before continuing its hunting activities. Secondly, in Hokkaidō waters, rough flensing of minke whale is permitted on board the whaling vessel, in part because there has been in recent years only one authorized land station in Hokkaidō (at Abashiri), and this allows the gunner to pursue other whales fairly soon after a successful hunt. Since minke whale meat requires prompt flensing to preserve its freshness, on-board flensing is essential to meet demand, though whalers point out that there is some trade-off in terms of shrinkage of the meat after on-board flensing occurs. The crew first winches the whale up onto the flensing deck situated in the stern of the boat, before the expert flenser (a land-based specialist from Honshū, who is added to the normal crew when boats operate in Hokkaidō waters) flenses the whale with the help of other members of the crew. The flensing operation usually goes as far as the second stage only, in which the 30 centimetre chunks of meat and blubber are prepared prior to finer cutting up into smaller blocks, which is later undertaken on land.

With the exception of minke taken in Hokkaidō waters, however, all whales caught in STCW must—according to law—be taken back to designated shore stations for flensing. The STCW flensing stations are generally smaller and simpler in layout than those used in LTCW. For instance, there are no boilers operating, since nowadays whale blubber is used for food and not for oil extraction.[10] Although some whaling operators, like those in Abashiri, may process some whale meat and blubber themselves in small workshops, most processing such as the salting and drying of meat, as well as the preparation of fertilizer and so on mentioned for LTCW, are carried out by other processors who specialize in such activities and who purchase their necessary raw materials either directly from whaling operators or through middlemen. The methods of sale vary from port to port and among different species.

The smallness of scale of the STCW landing stations thus gives rise to a structure of organization in which very few full time specialists are employed. The only experts in the processing of the whale are the chief flensers who separate the meat from the blubber, and cut the meat from the skeleton of the carcass. Other tasks may be

---

[10] The blubber of Baird's beaked whale used to be used for oil extraction in STCW, but this was stopped in the early 1980s for market and environmental reasons.

carried out by women and old people, who form the bulk of the work-force and are employed on a casual part-time basis, coming from the locality in which the station is found. Middlemen and distributors may lend a hand with flensing, and when necessary catcher boat crews may work on flensing as well. Not surprisingly, perhaps, the speed with which the work is conducted is slower than in pelagic whaling factory ships or on LTCW shore stations.

Similarities and Contrasts

It can be seen from the above description of large type coastal, pelagic, and small type coastal whaling (and, indeed, of net whaling as described in the previous chapter) that there are invariably two main sets of activities common to all types of whaling. One set relates to *catching* the whale; the other to *processing* it. Within each of these sets of activities there are certain similarities, as well as dissimilarities, among the four types of whaling described so far.

Let us start with the main dissimilarities in activities connected with catching. First of all, there is an obvious disparity between pre-modern net whaling and modern methods of whaling in the selection of hunting area, which was subject to local feudal government permission in the case of net whaling, but not so much in that of coastal whaling today; in the generally static nature of the search phase in net whaling, whereby whalers had to wait for the whales to pass by rather than actively go out and look for them; and in the use of several boats to drive the whale towards the net, as opposed to modern methods of using fast, powered boats equipped with harpoon guns.

Secondly, there are disparities among the various types of modern whaling methods. For example, in pelagic whaling a special vessel searched the area for whales and relayed information back to the factory ship, whereas in coastal whaling (both large and small) catcher boats worked totally independently. Moreover, whereas in STCW the echo sounder has been used as a device to bring the minke whale to the surface, in LTCW and pelagic whaling it was used to help the catcher boat keep track of all types of whales. Thirdly, in pelagic whaling, specialized boats were used to collect whales and take them to the processing unit, whereas in coastal whaling, the catcher boat itself performed this task.[11] Finally, whereas STCW can be categorized as "anticipatory" whaling—in that whalers wait for whales to come close to the coast in order to catch them—pelagic whaling (and to a lesser extent LTCW) is "dynamic", in the sense that whalers go out for long periods of time in active search of whales.

This point brings us to a discussion of the similarities to be found in the catching set of activities. Many of these will have been apparent from our description of the four types of whaling outlined here, but it should be stressed that in all types we can break down the set of catching activities into five distinct phases: hunting ground selection, search, chase, killing, and securement.

In deciding on hunting grounds, the gunner in the case of LTCW and STCW, or the commander of the fleet in pelagic whaling, relies on his knowledge of whale behavior accumulated through long experience, and on information obtained from recent hunts or supplied by other vessels in the same waters. Considerations of fuel consumption and time factors enter into his decision about where to hunt.

Similarities in the search phase include use of look-outs and signals (which may be secret, as we have seen in pelagic whaling), the monitoring of natural environmental phenomena, and the identification of the species of whale sighted. The same abilities, which include good vision, concentration and keen senses, are considered to be essential qualities for someone wanting to become a good whaler in all types of whaling.

In the chase phase, we find close cooperation between the gunner (or harpooner in the case of pre-modern whaling) and his crew, on the one hand, and the supreme authority of the gunner on board his vessel and the prestige accruing to his position, on the other. In addition, the gunner must have a good knowledge of the whale's behaviour in order to anticipate its movements and reactions.

We find that in the killing phase, the skill of the gunner in bringing his vessel to within range of the whale, his accuracy in shooting it, and the timing of the shot itself are all essential. All types of whaling make use of harpoons equipped with ropes. The fact that one harpoon was frequently insufficient to kill a whale (particularly in STCW before the introduction of the exploding harpoon) has meant

---

[11] The use of specialized craft to tow the dead whale to be processed was also found in net whaling in pre-modern times.

that new gunners could be trained by allowing them to fire shots
subsequent to the initial hit.

Finally, in the securement phase, the whale has to be prevented
from sinking, either by being pumped with air, or by having floats
attached to it, before it can be towed away to the processing unit.
Great care was taken in order to prevent damage or loss in quality to
the meat of the whale caused by waves and high water temperature
while the carcass was being towed back to the port.

In the processing set of activities, there are more similarities than
dissimilarities among types of whaling, and such disparities as exist
are fairly minor. On a shore station, for example, bones tend to have
been used for fertilizer and entrails for food, whereas on a floating
factory they have both been boiled to extract oil. Some land stations,
too, may subcontract part of the processing (bones, sinews, and
intestines) rather than carry out all aspects as on the factory ship.[12]
In other technological respects, however, the processing that takes
place in the factory ship and the other non-hunting ships of the pelagic
fleet is for all practical purposes identical to that encountered on the
shore station, so that the processing fleet can be conceptualized as a
floating shore station.

The use of the whale as such has changed very little over time.
With very few exceptions, the Japanese—unlike whalers in many other
nations—have sought to make total and variegated use of the whale.
In addition to the full utilization of red meat for food and blubber for
oil and food, the fact that various usages were found for the skin and
flukes (as salted food), the bones (as fertilizer or oil), entrails (as
fertilizer, oil or food) and for baleen, teeth and sinews (for craft
production) has affected the processing of the whale to such an extent
that this set of activities is very distinct from those found in other
whaling countries. This is true, moreover, for all the four types of
whaling described.[13] That the Japanese are very conscious of this

---

[12] We have noted that over time the new processing technique of freezing (introduced in the late 1930s) allowed a shift in preparation of whale meat from salting
to freezing. Moreover, changes in market demand have brought about a shift in the
use of blubber which is now mainly eaten rather than made into oil. These changes
apply, however, to all three types of modern whaling.

[13] It would be untrue to claim that all whale carcasses have always been 100 per
cent utilized by the Japanese. There have been periods when this has not been the
case. When LTCW started in the northeast, it took time to build up all the processing

difference can be seen in the way in which Oka Jūrō (the founder of one of the whaling companies [Tōyō Hogei]) as far back as the beginning of the century stressed that, when it came to Norwegian technology, Japan should adopt new technologies in the catching, but not in the processing, of the whale (Tōyō Hogei 1910).

The point to be made here is that, because each part of the whale is put to different uses and because various types of red meat are differently valued on the market, the various flensing activities have to be done with great care. This means that the early stages of flensing require great skill and that local casual day labour can only be employed, if at all, for the later stages of processing. We should note that all those employed in the hunting and processing of whales in whatever form of whaling, received payment partly in kind. This led to the development of a unique system of gift exchange and to a rich local food culture based on whale meat, which will be explored in Chapter 7. Before that, however, we need to look at another remarkably persistent feature of whaling in Japan—the recruitment of personnel—since it is this which acts as a bridging mechanism between the three types of modern whaling that we have outlined here.

---

facilities and whale bones, in particular, were wasted and allowed to pollute the shore. Later, in the days of the "Whaling Olympics" when there was fierce competition to catch as many whales as possible before the total quota was reached, there was not always time to process the carcasses fully before new whales were hoisted on board the factory ships, and the least valuable parts of the carcasses were dumped. The usage of whales as fenders between factory ships and catcher boats at this time—something all the whaling nations did in the Antarctic—also testifies to a less-than-maximum utilization of the whales.

Recently, bones of small type whales have also been wasted for the simple reason that, with the closure of LTCW, there are no processing facilities left to handle the limited volume of bones brought in by the STCW boats. The exception is a factory in Narita, which takes bones from Baird's beaked whales landed in Wadaura.

CHAPTER 6

Recruitment and Career Patterns

Our discussion of the work organization involved in whaling leads us to make use of the concept of a "whaling culture", based on the two quite separate activities of hunting and processing whales. The introduction of this phrase automatically raises the questions: how is a whaling culture in fact maintained and what is it that enables Japanese whaling culture to become "integrated"? After all, in the previous chapter, we have pointed to the existence of three types of whaling, as well as to two distinct sub-cultures based on hunting and processing. How are these different types and apparently separate sub-cultures linked? It is questions like these that we will address during the remaining part of this book.

To some extent, of course, the overall industrial structure of whaling in Japan helps bring about integration (a point to which we shall return in Chapter 8). Then there is the utilization of the whale itself, together with the forms of technology used in their hunting and processing. But these in themselves are insuffici :nt to enable us to talk about an integrated whaling culture as such. Rather, we must address initially the problem of how a whaling culture is maintained. In this chapter we will look at how the bearers of the whaling culture, the whalers, are recruited into this industry. By doing this, we will be able to show that recruitment patterns have, on the one hand, worked to slow down the dissemination of a whaling culture to new regions but, on the other hand, helped bring about an integrated whaling culture, in that they have served as linkages between the past and the present, between the three forms of whaling, and between the various whaling communities.

Whaling activities can, as we have argued in the previous chapter, be divided into two sets of activities; the catching of the whale and the processing of its carcass. Both sets of activities require great skills which are not learnt easily. It takes many years to be a good gunner or a skilful master flenser. Before the formal training of a whaler can begin, however, it is of great value for an apprentice to have been brought up in a whaling environment, where he can acquire a basic knowledge and understanding of shared community beliefs, technical skills, and so on. Not surprisingly, we find that most whalers have been brought up in such whaling communities. The great importance of personal connections when applying for jobs in Japan—a theme stressed by many of the informants—works in the same direction: companies tended to recruit whalers from a limited area, which then gave most of those companies a strong regional base.

At the same time, there has always been great geographical mobility in whaling operations in Japan. When whaling companies faced the problem of how to secure a skilled work force, as well as simultaneously being able to move operations seasonally or to new whaling grounds, they hit on two solutions: (1) the creation of a highly mobile work force; and (2) taking whalers from well-established whaling communities to train people at new locations. This is a pattern that was firmly established during the Tokugawa period, and we shall therefore start by briefly outlining some features found in the recruitment of whalers to the large net groups that hunted whales in those days.

Tokugawa Period Recruitment

One of the characteristics of Tokugawa whaling was its great geographical mobility. As new ways of whaling were developed, skilled whalers from one part of Japan were invited to other parts of the country by local authorities to exploit the whale resources there, as well as to serve as instructors for local people who wanted to take up whaling. Whaling entrepreneurs, moreover, travelled extensively in order to study the newest methods of whaling. For example, in the 1670s Fukazawa Gidayū from Ōmura (Nagasaki Prefecture) went both to Kayoi and Taiji in order to study their net whaling methods. He opted for the Taiji solution, which he brought back to Kyūshū where he first established a net whaling group at Uonome in Arikawa Bay.

But geographical mobility continued long after new methods of whaling were firmly established. Although some whaling groups— such as Wada's group in Taiji—worked the same whaling grounds for such as Wada's group in Taiji—worked the same whaling grounds for a century or more, this was by no means universal. Hunting rights rested with the local authorities, and licences to catch whales were usually granted for only a few years at a time. The authorities in Fukuoka, for example, issued licences for no more than three or five years, seldom renewing them once they had expired.[1] We have earlier seen that the situation in Arikawa was not much better.

This fluid situation was brought about by both economic and managerial problems. Whaling enterprises were among the largest, and structurally most complex, industrial organizations in Tokugawa Japan and whaling was thus a financial and managerial challenge to the entrepreneurs of the day (Kalland 1986). Many whaling enterprises failed even before they had successfully hunted their first whale. Others went bankrupt after just a season of whaling. But even the more successful operators—such as Nakao in Yobuko and Fukazawa in Ōmura—moved their operations frequently. And they took many of their workers with them, employing local day-labourers only for those tasks which required least skills.

The need for workers was great. Most whaling groups employed between 500 and 1,000 persons, who were about equally divided between hunting and processing operations.[2] These employees were hired as regular, seasonal or temporary workers. Only ten to fifteen persons were regularly employed, and these worked in management positions, often being relatives of the owners of the whaling enterprises. Those seasonally employed were mostly skilled workers

---

[1] Between 1766 and 1799 at least six operators (probably more since we have no information for the periods 1770-83 and 1794-98) successively worked the whaling grounds off Fukuoka's Ōshima. None of them renewed their licences. All the operators came from outside the Fukuoka Domain, i.e. from Karatsu, Yobuko and Ōmura (Kalland 1986:50).

[2] The Katsumoto group on Iki Island (Nagasaki Prefecture) employed 670 persons in the hunt in 1802, while 380 persons were so employed in the Ukushima group in 1817 (Hidemura 1952b:68; Kalland 1986:37). Similarly the Misaki group in Hirado (Nagasaki Prefecture) employed 587 persons in the hunt (Möbius 1893:1056). In 1815, the Arikawa group paid 314 hunters their salaries in advance (Hidemura 1952b:94). It was difficult to run a whaling group with fewer than 300 persons engaged in the hunt. The number of whalers on the land station fluctuated with catches. One Kyūshū group employed about 112 persons for the season, while 200-370 were taken on as day-labourers when whales were brought in (Hidemura 1952b:72-73; Kalland 1986:40). See also Appendix 3.

the most important whaling ground during this period, these experts were hired from a wide area, often from a number of different provinces. For example, most of the net-makers (*ami-daiku*), who doubled as crew on the net boats, came from the Inland Sea area, particularly from villages in Yamaguchi and Hiroshima prefectures (Hidemura 1952b:88; Hidemura and Fujimoto 1978:167). A number of persons working at the shore station were engaged for the season. They included some managerial staff, master flensers, boat builders, smiths, coopers, cooks and people who worked with sinews.[3]

Whalers engaged in the hunt were also hired from a wide area and boats were crewed by people from a large number of villages. It is known, for example, that the Misaki net group (in Hirado) employed crews for its 33 boats from 22 hamlets, while the crews of the 44 whaling boats in the Katsumoto group (Iki Island) were recruited from 21 hamlets. Both groups also manned some, or all, of their net boats with people from the Inland Sea area (Hidemura 1952b:76). Hamlets located on Ukushima provided more than half the boats to these two groups.[4] It is also worth noting that only one boat from Misaki itself and none from Katsumoto seems to have participated in these groups in which, as a whole, few local people were seasonally employed. As a result, whalers from a wide area were brought together, thereby providing fertile ground for cultural exchange.

Usually crew-members on one boat came from the same village and it seems likely that it was the task of the harpooner to hire the rest of the crew. The harpooners were thus intermediaries between the owners of the whaling enterprises and ordinary crew members. The harpooner and his crew made up a well-trained unit which was an asset to any whaling group. It was therefore difficult for newcomers from other villages to get accepted and whaling tended to become a special trade for certain villages—a tendency found with a number of other crafts in Japan.

Finally, there were the day labourers—both men and

---

[3] The division between regular, temporary and seasonal workers is still practiced in large corporations in Japan, and gives rise to the so called "dual structure" of the economy and the distinction between Japanese "lifetime" and "market-oriented" employment systems (see Dore 1973:264-279 in particular). For a general discussion on mobility and recruitment in present day companies, see Clark 1979, Chapter 5.

[4] Both these groups were owned by Masutomi Matazaemon, who was one of the largest—if not the largest—organizer of whaling in Japan between 1750 and 1850.

women—who were hired when whales were caught. As these people were called up at short notice, most of them were fishermen and farmers living in villages in the vicinity. Whaling was in southwestern Japan mostly conducted during winter, in other words in the agricultural off-season, and many farmers found it attractive to work in the whaling business since they earned much needed cash.[5]

Two main features stand out from this brief treatment of the pattern of recruitment during the Tokugawa period. Firstly, there was a sharp contrast between skilled and unskilled workers in terms of recruitment. The skilled workers, who were first of all found as crews on the whaling boats, and also as both master flensers and workers with sinews on the shore stations, were hired from a wide geographical area, whereas the unskilled labourers were hired locally. Secondly, skilled whalers from a number of villages worked together in the same group, often far from home. We shall see that the same pattern repeated itself in modern Japanese whaling.

Frame 5: Capitalization of Net-Whaling Groups

Capital both for investment and running a whaling group was extensive. 20-30 boats of various kinds had to be built and at least two nets prepared—not to mention the investment needed to build a shore station, with its working sheds, storage houses, living quarters, harbour facilities, tools, and so on. Running expenses included salaries to all the hunters, as well as to those who worked on the shore station. Many of the whalers were employed by the season, thus placing a substantial financial burden on any whaling enterprise which failed in the hunt. Other running expenses included taxes to the authorities, fuel, compensation to the host village for inconveniences caused to fishing activities, and so on.

Few people had enough capital to invest so heavily, and whaling groups consequently came to rely on loans. The operators borrowed from individuals within their own domains, from big merchants in Ōsaka and from feudal authorities (Fujimoto 1964). The creditors got

---

[5] So attractive in fact, that the feudal lord of Hirado issued an order in as early as 1626 that farmers were allowed to go whaling only if they had completed all farming work and had the spare time (Noma 1973:49).

whale products in return.

In cash terms, meat was the most valuable income, but residual products, which were used as fertilizer, were also important. Thus, officials from Kagoshima Domain participated in auctions held in the northern Kyūshū area so that they could buy fertilizer for their tobacco fields in Kagoshima (Hidemura and Fujimoto 1978:167). Kagoshima Domain also invested in a net group in Chōshū (Yamaguchi Prefecture) in order to gain access to fertilizer.

But it was whale-oil which was the most important product so far as feudal authorities were concerned. Oil came into great demand after it was discovered during a famine in 1732-33 that it could be used as an efficient insecticide in the rice fields, so the authorities did much to secure a steady supply of this product for their domains. Karatsu, Tsushima and Arikawa all ran public whaling enterprises at various times, while Hirado levied a special tax payable in oil, or forced the groups to sell oil to the authorities at prices lower than the market price (Fujimoto 1964:260). Other authorities extended favourable loans in the hope of attracting whaling operations to their own domains, or invested money in groups which operated in other domains.

Migrant Workers Versus Local Workers

Once modern technology was adopted and steam ships and harpoon guns began to be used, the range of operations was considerably enlarged, so that soon the whole of Japan, Korea, Taiwan and the Kuril Islands came within reach of the whaling companies. With the opening of these new grounds whaling became a year-round activity, with the whaling fleet hunting whales in southwestern Japan, Korea, and Taiwan during the winters and moving northwards during the summer months. These whaling operations thus had to be even more mobile than before. Moreover, there was little skill available locally near the new whaling grounds, so that the modern whaling companies faced the old dilemma in an even more acute form. How could they secure skilled workers for their operations?

One solution to this problem was to bring whalers from the old whaling communities in southwestern Japan and let them move with the seasons. Whaling thus became a year-round activity for the whalers from this region, rather than a by-employment during the slack agricultural season, which had often been the case before. This affected agriculture adversely in that it was left to wives and aged parents to take care of the land.[6] Consequently much farm land went out of use (Kalland 1989:99, 122) and villagers became more dependent on whaling than they had ever been before.

It was easy for the whaling companies to bring in people from the south. Most of the early whaling companies using the Norwegian method were established by entrepreneurs who came from old whaling communities in the southwest. Given the importance of personal connections when applying for jobs in Japan, these companies attained a strong regional base. A typical example is Tōyō Hogei, which was established when two important companies merged in 1909: Nagasaki Hogei which had a strong representation in Nagasaki Prefecture, thanks to its president Hara Shinichi who came from Arikawa; and Tōyō Gyogyō, whose base was in Yamaguchi Prefecture because of its president Oka Jūrō. Whalers from these two prefectures have continued to be well represented in Nissui, Tōyō Hogei's successor.

In Ayukawa, all the companies which established stations during the first decade of this century had this kind of strong regional orientation, as can also be seen from the titles of three of the companies which make use of geographical names (Table 1).

Table 1: The Early Whaling Companies with Stations in Ayukawa

Year	Name of company	Headquarters	Main recruitment area
1906	Tōyō Gyogyō	Shimonoseki	North-Kyūshū, Yamaguchi
1907	Tosa Hogei	Kōchi Prefecture	Tosa Prov (now Kōchi Pref.)
1908	Kii Suisan	Wakayama Pref.	Kumano Coast (incl. Taiji)
1911	Nagato Hogei	Senzaki	Yamaguchi Prefecture

Whalers from whaling centers in southwestern Japan—particularly from the prefectures of Wakayama (i.e. Taiji), Yamaguchi (the Nagato Coast), Nagasaki (particularly the Gotō Archipelago, including

---

[6] This kind of agriculture has since the 1960s become the common pattern in Japan and has been termed *sanchan-nōgyō*, or agriculture done by the three *"chan"*, i.e. "grandma", "grandpa", and "mom". As more housewives have found employment outside agriculture, it has recently developed into what is known as *"nichan-nōgyō"*, agriculture by "grandma" and "grandpa".

Coast), Nagasaki (particularly the Gotō Archipelago, including Arikawa and Ukushima) and Kōchi—were not only sent by their companies to the newly opened stations in northern Japan, but also to shore stations in Japanese occupied territories. Before World War II most of the whalers spent the summer months in the north (Tōhoku, Hokkaidō, Etorofu or the Kuril Islands), while the winter months were spent whaling off Korea, Taiwan and Ogasawara, or else farming at home. From the mid-1930s, however, Antarctic whaling became the most important activity during the winter.

With the establishment of new shore stations in Japan, local villagers were exposed to a new way of life, but despite the great demand for workers they were only slowly recruited into the whaling activities. So far as the companies were concerned, of course, local workers were cheaper than outside whalers who had to be brought in and often supported in dormitories at extra cost. But this expense was in many cases more than offset by the higher productivity of the trained whalers from the well-established whaling communities. Moreover, whaling companies were reluctant to hire people without connections who could vouchsafe for their good behaviour, and without such connections it was very difficult to get into the family-like enterprise that was typical of a number of whaling companies in the early stages of their development.[7]

This was particularly true of hunting operations where a well-trained and coordinated crew was a great asset. In the old days when nets were used, it was—as we have already seen—commonly the responsibility of the harpooner (*hazashi*) to recruit the crew on a boat, and frequently crew members all came from one village. The harpooners were the ultimate source of authority on the whaling boats and the success or failure of the whaling operations rested very much on them. The same has been true for gunners in modern times, and until recently it was also common for a gunner to recruit many of the crew members on the catcher boats, particularly in STCW. For example, when the father of one of the owners in Abashiri bought his first STCW boat in 1953, most of the crew came from Kagoshima

---

[7] In post-war Japan, connections have been created between whaling companies and institutions of higher education and the latter has recently been more responsible for the supply of crews on factory ships and, to a lesser extent, catcher boat than in the past.

Prefecture—the gunner's own home region.[8] On the other Abashiri boat several of the crew members were recruited through the late gunner and his wife. Gunners still tend to have more influence over recruitment procedures than boat owners, and as a result the recruitment base has been narrowed considerably.

Even today, there is a clear difference in residential origins between those employed on catcher boats and those employed at flensing stations. For instance, of the 44 men working on Nihon Hogei's two catcher boats in Ayukawa until they were laid off in 1987, only six persons actually lived in Oshika Town, of which Ayukawa is a part. Seventeen others lived in the prefecture, whereas almost half—21 whalers—lived in other prefectures. Of these, thirteen lived in Taiji. The situation among those who worked at the flensing station, on the other hand, was quite the opposite since more than half lived in Oshika Town and only three persons (10.7 per cent) lived outside the prefecture (Appendix 4b). This is the same situation as that which prevailed during the Tokugawa period when those working on the shore stations tended to be recruited from farming and fishing villages nearby, whereas the harpooners, their crews and the crews on the net boats were recruited from a much wider area.

We have already remarked that on the shore stations, the initial flensing was an operation which required high skills and exact timing, factors that were less crucial in the later stages of processing. Local people were therefore first employed at the shore stations where they did the less demanding tasks, leaving those that required great skills— such as the initial flensing—to a group of skilled migrant whalers. This is a strategy that the STCW companies have also employed. Gaibō Hogei, for example, brings its master flensers down from Ayukawa to Wadaura during the season there, and in the past when one of the Abashiri boats operated off Nemuro, the master flenser went from Abashiri to Nemuro in order to do the main flensing there. Until the moratorium went into effect, the STCW boats took master flensers from Ayukawa on board the boats when they caught minke whales off Hokkaidō.

---

[8] There are no descendents of these whalers in Abashiri today. Most of them are believed to have returned to Kagoshima in southern Kyūshū.

Plate 1: A manhole, Taiji.

Plate 2: Post office sign in Ayukawa

Plate 3: The Fire Station in Ayukawa

Plate 4: LTCW catcher boats in Ayukawa harbour.

Plate 5: Beaching of the Nihon Hogei catcher boat, *Toshi-maru No. 16*, in Ayukawa 1989.

Plate 6: Net whaling in the Tokugawa period (Taiji).

Plate 7: Catcher boat in the Antarctic

Plate 8: Leaving port for the Antarctic.

Plate 9: Factory ship approached by a catcher boat in the Antartic.

Plate 10: STCW catcher boat searching for a whale.

Plate 11: Striking a minke whale off Hokkaidō.

Plate 12: On-board flensing of minke whale (STCW, Abashiri)

Plate 13: Flensing a board factory ship in the Antarctic.

Plate 14: Flensing Baird's beaked whale, Ayukawa.

Plate 15: STCW flensing station, Abashiri.

Plate 16: Making dried and marinated *tare* whale meat, in Chikura nea
Wadaura.

Plate 17: Women scaling and packing dried *tare* whale meat, Chikura.

Plate 18: Japan bashing.

Plate 19: Whale memorial, Ukushima.

Plate 20: Lanterns commemorating the souls of whales - and donated by various whaling and community institutions - about to be floated out to sea. (Ayukawa)

Plate 21: Entrance to Kaidô shrine flanked by whale jaws, Arikawa.

Plate 22: Whale memorial service, Ayukawa.

Plate 23: Boys training for the Taiji whale dance.

Plate 24: *Meizaiten* festival, Arikawa.

Plate 25: *Ginzake* silver salmon in Ayukawa.

Plate 26: Whalers sharing a meal on board a STCW catcher boat, Abashiri.

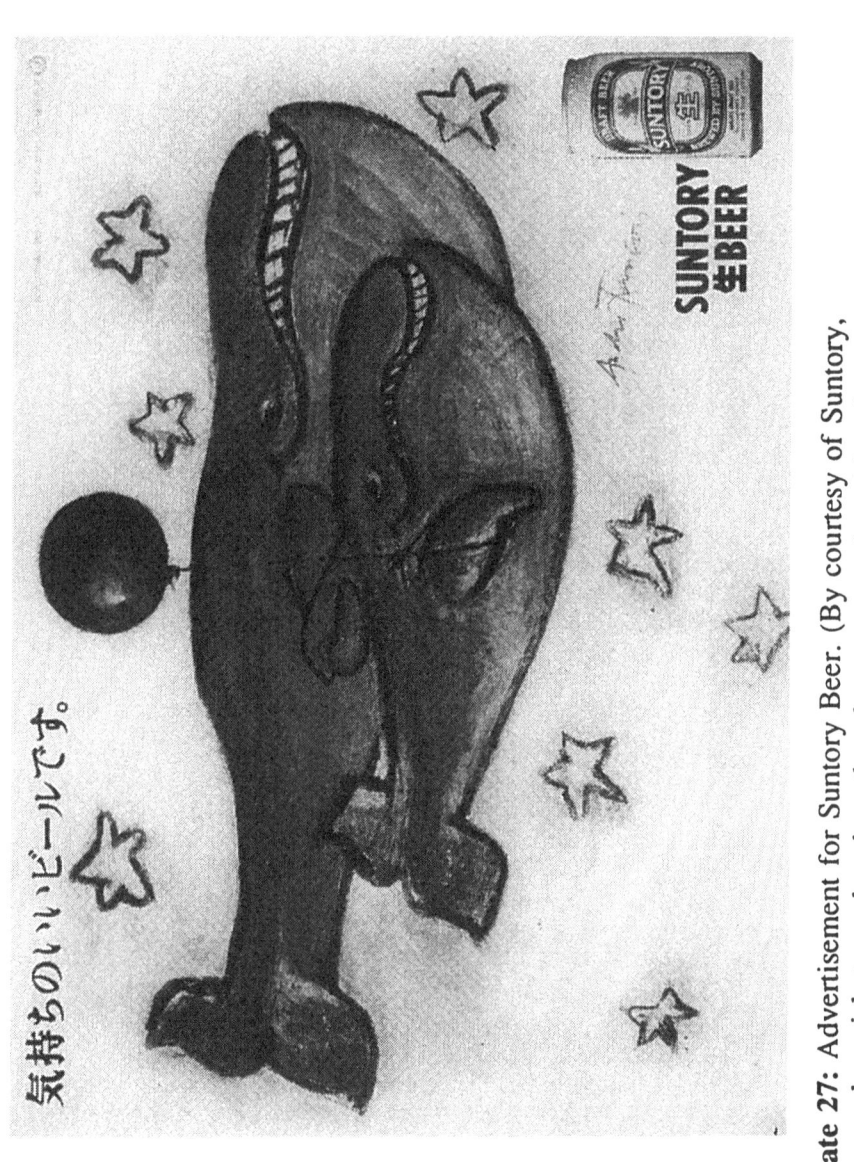

Plate 27: Advertisement for Suntory Beer. (By courtesy of Suntory, who wish to make clear that the company is neither for nor against whaling.)

Plate 28: A toy from Arikawa. A whale flanked by tow boats (*mosô-bune*) is mounted on a wooden board with wheels.

Permanent Resettlement

Given that there was a large number of migrant whalers working for whaling companies, it should come as no surprise to find that some of them settled in the north. Many whalers found their wives in the communities they visited and, although some brought their wives back to their home communities, others made their wives' communities their own. There were also other personal preferences and reasons for whalers from southwestern Japan settling throughout northern Honshū and Hokkaidō. Although it is somewhat difficult to get a good picture of migration patterns from whaling communities in the southwest to those in the north of Japan, and although it is even more difficult to quantify this migration, what we now wish to do is look at the whaling town of Ayukawa in order to see how its population expanded and changed through in-migration.

Ayukawa was nothing but a small fishing community before Tōyō Gyogyō established the first whaling station there in 1906. In 1887, for example, the village had only 332 inhabitants, a population which had increased to 3,660 by 1952. During the same period, the number of surname groups living there increased from 12 in 1879 to 67 in 1911, and finally to 261 in 1985 (Table 2). Although we cannot rule out the possibility that old Ayukawa residents adopted new surnames for reasons of convenience (such as trying to avoid military service), the explosion in the number of surname groups clearly testifies to considerable in-migration. In 1985, only 22.4 per cent of the households in Ayukawa used one of the original surnames represented in the village during the late nineteenth century, and this is considerably less than the percentages for the other hamlets of Oshika Town.[9]

What is the significance of these figures? In 1879, Ayukawa was still a remote fishing hamlet located at the tip of the Oshika peninsula. Its only claim to fame was as a stop-off point for visitors to the sacred island of Kinkazan, situated just off-shore. In 1906, however, the hamlet was chosen by Tōyō Gyogyō as a landing station for its LTCW operations in the north Pacific, and by 1911 nine of the twelve

---

[9] The percentages of the households that in the 1980s used one of the original surnames found in the respective hamlets are as follows: Ayukawa 22.4% (12 original surnames), Kugunari 62.1% (12), Kyūbun 76.6% (6), Kobuchi 69.6% (9), Ōhara 66.0% (16), Koamikura 75.4% (7), Niiyama 85.3% (5), Tomari 90.5% (6), Yagawa 63.0% (8), Oyagawa 84.4% (5), Samenoura 93.3% (2), Yoriisohama 69.4% (3 surnames) (Oshika-chō 1988).

Table 2: Ayukawa Surname Groups (1879-1985)

	1879	1911	1960	1985
Old surname households				
Original surname groups	12	12	12	12
H.holds with old surnames	52	82	186	181
as percentage	100.0	55.0	23.2	22.4
New surname households				
New surname groups	0	55	259	249
H.holds with new surnames	0	67	616	627
as percentage	0.0	45.0	76.8	77.6
Total no. of households	52	149	802	808

Source: Oshika-chō 1988:131-135.

whaling companies operating in Japan at that time had established stations in the area, four of them in Ayukawa hamlet itself. As we have already pointed out, the establishment of a landing station in Ayukawa led to the development of a large number of associated activities—not least of which was the production of whale oil—so that the hamlet rapidly became a thriving town, complete with dormitories for migrant workers, inns, and bars. But were did all these new inhabitants come from?[10] In the mid-80s, only 28 per cent of the household heads originally came from Ayukawa. The other households had been established in the village either as a result of whole households moving in, or more frequently by outsiders marrying local girls and settling in the town. These immigrants came from all over Japan (Appendix 5) and whereas people from Nagasaki, Kōchi, Waka-yama and Yamaguchi prefectures mostly came to Ayukawa in order to crew on catcher boats, to man the shore stations and to work as artisans in whale teeth and bones, people from Tōkyō and other places

---

[10] Fortunately, we have information about the geographical background of the household heads in Ayukawa, thanks to a dedicated local researcher who a few years ago interviewed most of the households in Ayukawa and cross-checked details with other sources. Although these figures cannot be as accurate as they would be if they had been based on the *koseki* household registers, they do give a clear picture of the pattern of migration into Ayukawa.

not formerly related to whaling activities resettled by establishing various forms of businesses supporting the whaling industry.

Whalers who married and settled in the north provided many local people with the connections they needed in order to get employed by the whaling companies. As those acquainted with Japanese society will be aware, a marriage is not simply an agreement made between two individuals, but involves their households, too. Thus, when a migrant whaler or other came, for example, to Ayukawa and married a local girl there, he immediately acquired a set of relatives who helped him settle down in local society and who, of course, relied on him to provide them with the necessary introductions to enable them to get employment in the whaling companies stationed in Ayukawa. In this way, of course, women acted as an important intermediary between seasonally migrating whalers and the local community.

Frame 6: Southern Whalers in the North

Even today whalers from the south play important roles in the whaling communities of north Japan. For example, three of the board members of the Ayukawa Old Boys' Association (*OB-kai*) of Nissui are from the south. One is the chairman whose father came from an old whaling community in Yamaguchi and was employed by Tōyō Gyogyō. He was sent to Ayukawa in 1907 where he met his wife, who then accompanied her husband to various shore stations until the first child was born. The other two are the vice-chairmen, who were born in Arikawa and Ukushima and have both married Ayukawa women. The secretary, too, was born in Niigata and adopted into Ayukawa as a child.

One of the vice-chairmen of the OB Association is also the present chairman of the Fishing Cooperative Association in Ayukawa. He was introduced to Nissui by his uncle (mother's brother) who worked as a gunner for that company and whose wife was from Ayukawa, and he himself married his uncle's niece-in-law. His wife's mother used to work for the same whaling company in which he himself is at present employed.

Of the three STCW boats registered in Ayukawa, one is owned by someone from Kumamoto Prefecture in Kyūshū, and has a captain from Yamaguchi. The owner was first employed by Taiyō Gyogyō through an introduction by his brother-in-law when he was 18 years old, and he worked both on refrigerator and factory ships before spending a couple of years in Ayukawa, partly to supervise Seiyō Hogei, a subsidiary of Taiyō engaged in STCW. In 1985 he resigned from Taiyō and bought Seiyō Hogei. He had moved his family from Kumamoto to Yokosuka in 1961, and in 1988 he brought his wife to Ayukawa. The manager of another STCW company, Nihon Kinkai, is from Yamaguchi.

There are three craftsmen producing artifacts from sperm whale teeth in Ayukawa, and all three are descendents of settlers from Kyūshū. The father of one came from Arikawa and worked for both Tōyō Hogei and Nissui as a flenser. The son followed in his father's footsteps but soon found it more interesting to sit on deck carving whale teeth. The other craftsmen are brothers born in Karatsu, an old whaling center in Saga Prefecture (see portrait of Nakamura Kanji in Chapter 3).

Two STCW are registered in Abashiri and both have important crew members whose family background is in the south. The captain of one of the boats hails from Kagawa Prefecture in Shikoku, while two brothers who work on the other boat (one of them as bosun) are sons of a gunner from Arikawa (see case study in this chapter).

Most, if not all, of these whalers keep in touch with their relatives in southern Japan. Some of them visit their relatives every year, and there are frequent, and continuous, exchanges of gifts between them. Letters, telephone calls, and greetings through third persons keep them surprisingly well informed about the activities of their relatives more than 1,000 km away. These settlers thus continue to bridge the gap between whaling companies and contribute in important ways to the formation of a national integrated whaling culture.

There were, moreover, other institutions—particularly strong in the northern Tōhoku region of Japan—which served to enmesh a newcomer into a local community. One of these was the extended household, or lineage, group generally known as the *dōzoku* (in Ayukawa called *make*), consisting of a main household (*honke*), branch households (*bunke*), branches of branches and so on, all "bound by mutually recognized links of genealogy" (Fukutake 1982:31; see also Chapter 7). In Ayukawa there seem to have been about half a dozen strong kinship groups in the past, each centered on an affluent *honke* that built its wealth on land holdings and on the ownership of large nets (*amimoto*). It was some of these *honke* which together established the whaling company Ayukawa Hogei in 1925.

Another institution was the relationship established between what were known as *ebisu-oya* (the "Ebisu parent", Ebisu being a deity particularly important to fishermen and merchants) and *ebisu-ko* (the "Ebisu child"). In the old days it was common to change one's name in connection with important *rites de passage*, and on marriage a man often received an *"ebisu"* name (*ebisu-na*) which was given to him as *ebisu-ko* by his *ebisu-oya*. Ideally, the relationship between the two men was very close, "almost like that of a child toward his father". The head of the *honke* often served as the *ebisu-oya* for members of his branch households (although the reverse could also occur), and net-owners also served as *ebisu-oya* for those working on the nets. More pertinently, heads of old Ayukawa households often served as *ebisu-oya* for the immigrant whalers who married local girls.

An important point to note here is that the relation between Ebisu parent and child was not between individuals, but households, so that 'members of the immediate family of the *ebisu-ko* also found themselves tied to the family of the *ebisu-oya*. Moreover, the Ebisu relationship ideally extended beyond the lifetimes of the individuals concerned. A marriage between a whaler from the south and an Ayukawa girl thus meant that the whaler became firmly integrated into the community, first by linking him to one of the existing kinship groups, and then through his *ebisu-oya* to an even wider social network. This network was, and still is, kept "warm" by an extensive exchange of gifts—in which whale meat until 1988 played a crucial role (see Chapter 7), and by mutual obligations to assist one other. One such obligation was to introduce the other to a third party, in other words, to the whaling company in which a man was employed.[11] A single marriage, therefore, provided a number of young men related to the bride with possible connections to the whaling companies.

Herein lies the importance of permanent settlement for the dissemination of whaling culture, in that it facilitated the recruitment of whalers from new communities. It did not put an end to migration of course, not even to that of the whaler who moved and established his family in the north, although resettlement reduced the extent of his

---

[11] In many respects, the Ebisu system in fishing villages in the northeast of Japan seems to have been used by net owners to secure a continuous supply of labour. In this, it differs from that found in a whaling community like Ayukawa where the function of the *ebisu-oya* was to act as intermediary between the outsider and community, rather than as employer.

migration since he usually settled in a community with a shore station from which LTCW (or STCW) was conducted part of the time.

Case Study: A Kinship Group From Arikawa

In order to illustrate better our argument with regard to the importance of mobility (both seasonal and permanent resettlement), the importance of connections in getting employment in a whaling company, and the role of the whaling company in recruitment, let us take a look at a patrilineally based kinship group in Arikawa.[12] This kinship group was originally based in Arikawa, but today has affiliated households as far north as Hokkaidō, thanks to migration and marriage with local families, and we can trace through it the importance of connections, the great geographical mobility of kin members, the overall flow of personnel among the various forms of whaling (i.e. STCW, LTCW and pelagic), the variety of jobs in whaling undertaken by members of a single kinship group, and the frequently close teacher-pupil relations that existed with regard to the training of new skills.

If we turn to Figure 1, we can see four generations of household heads represented in several different households, and referenced *A* to *M*, most of them heavily involved in whaling. *A* was not a whaler but a fisherman, although he may well have worked occasionally on the Arikawa flensing station. *A*'s younger sister's husband was, on the other hand, employed by Tōyō Gyogyō and worked on a catcher boat. Three brothers—*A*, *C*, and *D*—all got jobs through this uncle (*i*). The eldest of the brothers (*B*) who stayed on in Arikawa in order to take over his parents' house, became a flenser on a factory ship, whereas his younger brother (*C*) became a flenser on various shore stations throughout Japan. Eventually, he met and married a Hokkaidō woman and settled in Kiritappu. One of his sons lives in Abashiri (*J*) and two in Nemuro (*H* and *I*). All three have worked with their uncle (*D*), on STCW.[13]

---

[12] In this group all the household heads are related to each other through the male line (with one exception involving an adopted heir).

[13] Two other sons work for the Fishing Cooperative Association in Kiritappu and as a driving school teacher in Kushiro, respectively.

Figure 1: A Kinship Group from Arikawa

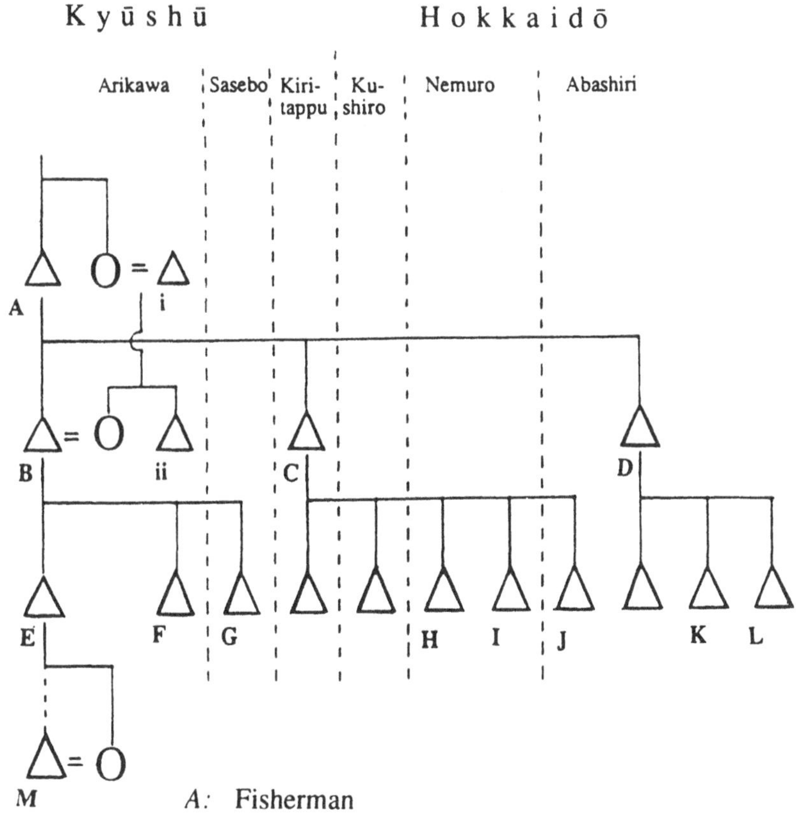
Kyūshū
Hokkaidō
Arikawa
Sasebo
Kiri-tappu
Ku-shiro
Nemuro
Abashiri
A
i
B
ii
C
D
E
F
G
H
I
J
K
L
M

A: Fisherman
B: Flenser (Nissui factory ship)
C: Flenser (Tōyō/Nissui shore station)
D: Gunner (Nissui LTCW/Pelagic --> STCW)
E: Gunner (Nissui --> STCW --> Taiyō)
F: Gunner (Nissui --> STCW --> Taiyō)
G: Flenser (Taiyō)
H: Deckhand (STCW)
I: Deckhand (STCW)
J: Cook (STCW)
K: Deckhand (STCW)
L: Bosun (STCW)
M: Wholesaler (Taiyō)

Legend: --- Adopted son
= Marriage

The youngest of the three brothers (*D*) was employed by Nissui and started as a cook under his uncle, going to the Antarctic three times before World War II. He was also sent to various stations in Japan and to Korea and the Kuril Islands. He met his wife in Hokkaidō and brought her to Arikawa just before the outbreak of the Pacific War. During the war Nissui bought five small catcher boats and *D* took employment on one of them, operated from Abashiri by Hōyō Hogei, a subsidiary of Nissui. After the war he preferred to continue in STCW, and so moved his family from Arikawa to Abashiri. He worked as gunner on several boats, some of which also operated from Ayukawa and Fukui, before taking employment on one of the boats at present stationed in Abashiri. Two of his sons (*K* and *L*) took employment on the same boat, and the youngest was bosun until he was laid off in 1988. He fervently desires to follow in his father's footsteps and become a gunner.

The brothers *E*, *F*, and *G* are sons of a flenser, but apparently it was their mother's brother (*ii*), rather than their father, who was most influential in getting them into whaling. This uncle worked on a catcher boat and greatly influenced his nephews. Although the youngest brother (*G*) became a flenser for Taiyō, the two eldest became gunners for the same company.

F started his career in Nissui as a "boy", but quickly displayed his talent in spotting whales. After six years with Nissui, his uncle (*D*) invited him to Abashiri in order to train him as a gunner. After being apprenticed for about two years, he became the gunner on a STCW boat in Ayukawa where he was working. Then he was offered a job with Taiyō. Before he left the Ayukawa company, however, he trained a new gunner, who was originally from Arikawa but was later killed in an accident when he was loading a harpoon.[14] After eight years all told in STCW, *F* became a gunner with Taiyō, a position he held until his retirement fifteen years later, when he opened a fish shop and a souvenir shop in Arikawa with his savings and the retirement money that he had received from his company.

His elder brother (*E*) had a similar career, starting in Nissui before being trained as a gunner in STCW. He was recruited by Taiyō three years before *F* and likewise trained his successor as gunner on the STCW boat, a person who was born in Arikawa but who has now

---

[14] His wife, who was also from Arikawa, returned to her natal home with the children after the accident.

settled in Shiogama City in Miyagi Prefecture. *E* met a tragic death in the Antarctic, however, after he was thrown down from the harpoon platform onto the deck by a large wave which hit the boat. He left a daughter and a 39 year old widow, whom people from her husband's company helped set up a whale wholesaling business. Today she is the only wholesaler engaged to any significant degree in the distribution of whale meat in Arikawa, and the business is now run by the adopted husband of *E*'s daughter (*M*). In his small workshop he processes the whale meat (including blubber) into a range of products specially cherished by the people of Arikawa.

Members of this kinship group have contributed to the development and spread of whaling culture in important ways. *D*'s widow remembers how her husband taught her Arikawa rituals related to whaling and took great care in religious matters, always offering some of the whale's tail to the boat altar (*kamidana*)—an old tradition in the south. Several of the members have trained new whalers both in Abashiri and in Ayukawa, and by switching between STCW, LTCW and pelagic whaling they have contributed to the formation of an integrated whaling culture encompassing all types of whaling. Back in Arikawa the head of the main house (*honke*) has moved into the wholesaling and processing of whale meat, thereby developing new knowledge locally, while his uncle (*F*) runs the largest retail shop for whale meat in the township. Both prepare dishes for hotels and inns and both are thus caretakers of a rich culinary tradition (see Chapter 7).

Career Patterns

The above example also tells us a lot about career patterns in Japanese whaling companies. We have seen that members of this kinship group have worked in all phases of whaling, but nobody has switched from one to the other. There is, and has probably always been, a sharp division between the career paths of the two sets of activities relating to whaling, so that a whaler is either involved in hunting the whale, or in processing its carcass. Crossing this line has indeed been rare. On the other hand, we have seen that several members did switch between LTCW and STCW.[15] As we shall see shortly, a pattern can be seen in this kind of mobility between types of whaling.

For those employed on the catcher boats, the first step in a career was almost invariably to be taken on as cook (in STCW) or as a "boy" (in LTCW). Being the youngest on board could not have been easy, particularly in a society where seniority counts for as much as it does in Japan. It was thus a period when the whaling apprentice would try to prove his worth. Some quit at this stage—especially those not able to get up early in the morning to prepare the crew's breakfast—but most stayed on board and became deckhands. In the 1950s when whaling expanded, one could climb the promotion ladder rather quickly since whalers were in great demand. The ultimate goal for most of them was to make it to gunner. Before anyone could hope to aspire to such a position, however, he most likely needed to excel as deckhand, quartermaster or bosun, and had to display leadership abilities. Only after many years' hard work could he hope to be allowed to try his hand at harpooning, as an assistant to the gunner, and shoot the second—and usually fatal—shot. A training period typically lasted a couple of years or more. Some large whaling companies, such as Taiyō, had a special training field outside Ayukawa where apprentices could practice by shooting at moving objects on a revolving platform. Some STCW companies had their apprentices shoot at floats and the like on days when the weather was too bad for whaling. For both STCW and LTCW, however, most of the training was carried out during hunts, since anticipating a whale's next move is not something that can be learnt from books, but only through careful observation. Shooting at floats in calm water was also completely different, of course, from shooting at a whale in stormy weather in the Antarctic.

In the Arikawa kinship group, two brothers (*E* and *F*) both started their careers on Nissui's catcher boats, but later switched to STCW boats in order to be trained as gunners. When pelagic whaling expanded during the 1950s there was wide-spread head-hunting—of gunners in particular. Large whaling companies like Nissui, Taiyō and Kyokuyō could offer much higher salaries than the STCW companies, and it was also more prestigious to be employed by these companies. As experienced gunners moved from STCW to LTCW, many would-

---

[15] LTCW in this context includes pelagic whaling because these two types of whaling with few exceptions involved the same personnel.

be gunners sought employment with the STCW companies, so that STCW came to serve as a training ground for gunners and the overall efficiency of these boats declined. Owners tried to raise this issue with the big whaling companies, but to no avail. The latter also took over bosuns and engineers from the STCW boats, but it was the gunners who were most difficult to replace. More recently, however, the process has been reversed for, with the contraction of pelagic whaling and LTCW, some laid-off whalers have found employment in STCW.

There were other positions to be filled on the catcher boat—for example, those of engineer and captain. Since both needed special skills, they had a somewhat different career path from others on the boat. The captain—who in STCW sometimes doubled as gunner—needed a seaman's licence, and Nissui had its own maritime school in Kyūshū. There are a number of different certificates, each giving the holder the right to navigate a ship of a certain class, and the highest of these is needed for someone to be captain of a factory ship, or commander-in-chief of a whole fleet. The chief engineer has other skills which require him to be an expert on ship engines—a task for which he was specially trained. In a way, he has been a bit alien from the rest of the crew in that the engine room provided a career path in itself, regardless of whether it was located on a whaling vessel, or not.

At the LTCW station or on the factory ship, a man's career usually started with his being a "boy". In his mid-teens a boy was made to do various odd jobs, either in the processing itself or in the office, before being promoted to work in the kitchen. Only after a couple of years, if he was lucky, would he be allowed to do "real work" with the whales. While young, he could be sent to various departments, to the flensing deck, for example, or to the boilers in the oil extraction section, or to the freezing section. But before long, he would be firmly established within one of these sections, and would begin his specialization. In the oil extraction section he could become an expert either on the Hartmann or on the Kværner boiler. On the flensing deck, the young man would first work as a *kagihiki*, "wire puller", helping the flensers to make accurate cuts before, after a couple of years more, being allowed to handle knives. To become a master flenser was the goal, but many failed to get that far.

The career pattern on STCW shore stations has been quite different from that of large type whaling. Much of the butchering is done by semi-skilled workers, women and retired whalers who give a hand when whales are landed. There is thus no long-term career path in the way that we find on factory ships and LTCW stations. Nevertheless, there are a few master flensers who have started out as assistants, before being taken on by the large whaling companies in the heyday of whaling. However, this kind of head-hunting did not have the same ill effects as the take-over of skilled gunners, since there were some flensers who had retired from the demanding work in the Antarctic, but who were still fit to flense a minke whale on a shore station at home. Thus one career pattern was to start out in STCW, switch to LTCW for several years, before finally returning to STCW. With the reduction in whaling operations, more LTCW flensers tried to get employment in the STCW companies.

We have seen that there are two distinct career paths in whaling—one for hunting whales and another for the processing of the carcass. There is almost no crossing of personnel from one path to the other. In some types of whaling—particularly in pelagic whaling—the two sets of personnel have almost no contacts at all during the whaling season, for the catcher boats returned to their factory ships only every two weeks or so, in order to stock up with provisions. Contacts were more frequent on the shore stations were people had more opportunities to meet—particularly in STCW where hunters and processors met on the wharf daily, and where hunters might even lend a hand in flensing. Contacts were especially strong during the minke whaling off Hokkaidō, when master flensers from the shore stations performed the main flensing on board the catcher boats with the help of the crew.

So, although people did not switch from one set of activities to another, it can be seen that there was wide-spread mobility within each of the two sets of activities. Employees climbed the ladder of seniority from "greenhorn" to positions of authority, with promotion depending on one's own talent and ability to work hard. Although much of this mobility occurred within the same company, there are sufficient examples of switching between types of whaling to warrant the claim that recruitment and career patterns served to break down the barriers between the three types of whaling.

This mobility, of course, to some extent facilitated a diffusion of knowledge between the various forms of whaling. However, other institutions were needed to break down the barriers between the two sets of activities relating to whaling and we have, in our example from Arikawa, seen how one kinship group could be such an institution. By having whalers engaged in both sets of activities, kinship relations could thus be an efficient way of disseminating knowledge. Two other institutions which did much to break down barriers were the whaling communities themselves and the whaling companies. In the following two chapters, therefore, we shall take a closer look at these institutions.

CHAPTER 7

Local Whaling Culture

In the previous chapter, we looked at the way in which whalers have been recruited, both in traditional and more modern times, and showed one way in which a whaling culture begins to be formed. However, our discussion has focussed on the differences found among those employed in the hunting and processing of whales, in that we have seen how the two groups tend to recruit members along separate lines, so that cross-mobility between these two separate sets of activities has been infrequent—except to some extent in the case of STCW. Given the separation of personnel brought about by specialization in hunting and processing, we now need to ask whether there are mechanisms involved in whaling which in fact serve to link people—in other words, whether there are not some integrating mechanisms based on similarities, rather than on differences.

This chapter, then, focusses on how the local community has adapted to whaling and shows how the divisions of hunting and processing are in fact overcome at the village level, where community identity is created through a number of means. These include the way in which the local village has benefitted from the establishment of shore stations by whaling companies, the development of non-commercial distributions of whale meat, the evaluation of whale meat as cuisine, and the rituals and beliefs connected with whaling that have developed around the local community shrines and temples. Our argument will be that such identity creation gives rise to a *local* integrated whaling culture at the village level. Just how each local integrated culture is then fused into a *national* integrated whaling culture will be the subject of the following chapter.

The Japanese Village

Before proceeding to discuss the way in which whaling communities as such have come into existence, we need to look at the salient features of the Japanese village (*buraku*) in general, since no discussion of the whaling industry can ignore the structure of rural society in Japan. We will use the concept "village" to denote a sub-unit of a larger administrative entity, such as *mura* ("administrative village"), *machi* ("town") or *shi* ("city"). Most of these were administrative units in the Tokugawa Period and the village "survives" today as a "ward" (*ku, chiku, ōaza*) within the larger administrative unit. The village has its own elected officials, such as headman (*kuchō*) and accountant (*kaikei*) who are responsible to all member households of the village, which is itself further divided into neighbourhood groups (*tonarigumi*). The village assembly is usually made up of the elected officials and heads of the neighbourhood groups. Its authority covers such village affairs as the administration of common property (shrines, cemeteries, land, community centre) and the arrangement of annual events (festivals and sport days). The village assembly is also an important pressure group *vis-á-vis* local government in that it is expected to suggest projects which will benefit the villagers (building roads, sportsfields, children's play-grounds, schools, and so on). One other important feature of the village is that this is the level at which most of the local associations—such as those for children, school-children's parents, youth, fire brigade, housewives, and old people—are based.[1]

Briefly, we may say that a village consists of a number of house-holds located in close proximity to one another, and that it is the household—and not the individual—which is the basic unit of administration. Each household consists of anything up to four generations of people who are, for the most part, patrilineally related since the line of immediate descent is usually (though not necessarily)

---

[1] There is a large body of anthropological literature in English on the Japanese village so defined. See for example, Embree (1939); Norbeck (1954); Beardsley et al. (1969); Dore (1978); Smith (1978); Moeran (1984). For "villages" in urban settings, see Dore (1958); Vogel (1963); and Bestor (1989a).

through the eldest son.[2] Other sons (and, occasionally, daughters) may often establish what are known as "branch" households (*bunke*), and because these receive land and/or property from the "main" house (*honke*), the main and branch houses tend to stand in an economic relationship, in which the branch household gives labour in exchange for the initial gift of property. Since a branch house can itself form branch households of its own, and since the original main house may form a number of branch households over time, it is possible—particularly in the northeastern regions of Japan (Fukutake 1949, 1956; Isoda 1951; Dore 1959:364-6)—for several households in a village to form a lineage (*dōzoku*) based on the exchange of property and labour. In other parts of Japan—particularly in the southwest—where the economic relation between main and branch households is not so strong, ties are nevertheless to some extent maintained between households at the ideological level through common ancestor "worship" (Plath 1964; Smith 1974).

In many Japanese villages a few main households are intimately connected with the village past. They are the "founding fathers" of the village, and are often believed—as in Ayukawa, Taiji and Arikawa—to descend from *samurai* who for various reasons chose to settle in the countryside. In the past they tended to monopolize key positions in the settlement, such as those of headman, village elders, and members of the shrine committee (*miyaza, miya-sōdaikai*). Their interest in controlling the shrine committee reflects the traditional importance of the local Shintō shrine in Japan, for it is here—where the tutelary deity is enshrined—that to a great extent the village as a unit is defined. Most, if not all, the main village festivals are centered on the shrine, and through these festivals the village hierarchy is reinforced and the common village heritage stressed.

At the same time, although the structure of the household in Japanese rural society is based on the administration of privately owned land, the village itself has been concerned with the administration of common resources—forest land (which yielded, for example, timber for fuel, grass for green fertilizer or animal fodder, and reeds for thatching house roofs); water for irrigating rice paddy

---

[2] In Ayukawa, for example, the line of descent has been through the eldest child, thus encouraging the practice of adoption of male heirs through marriage. In the Gotō Archipelago descent has, on the other hand, been through the youngest son (Naitō 1970).

owned by individual households; fishing grounds; and the shore-line itself (for pulling up nets and boats, and for drying nets and fish). Since one or more of these common resources is vital to agricultural or fishing villages respectively, it can be said that the Japanese village is in effect an administrative unit, concerned with the management of natural resources.

In coastal villages the administration of the common fishing territory has, through the Fishery Laws of 1901 and 1949, been transferred from the village to local fishing cooperative associations, or FCAs (Ruddle 1987; Kalland 1981:135-136; 1990). Only members of the local FCA are usually allowed to exploit the resources within the FCA's exclusive territorial waters, and only active fishermen residing within the borders of the fishing village, are—in theory at least—entitled to become members of the FCA. This close connection between the fishermen and the administrators of the fishing territory is one of the main reasons why management of coastal resources has been so successful, at least compared to the management regimes found in most other countries (Kalland 1990).

The Local Community and Whaling

Given that the rural or fishing community has been an independent, self-governing entity which managed its own natural resources, Japanese whaling companies found themselves automatically obliged to negotiate with those villages in which they wished to establish shore stations. They had to pay to gain access to natural resources controlled by the villages and such payments were seen by villagers to be a form of "compensation" for inconveniences caused by the activities of whaling boats to the local fisheries, as well as for the damage to the environment caused by the whaling companies in operating shore stations, where flensing polluted coastal waters, bones the beaches, and the boiling of blubber the atmosphere.

The payment of compensation by whaling operators is not new. During the Tokugawa Period, whaling groups used to pay annual fees to the villages from which they operated, and these fees were expressly to compensate local fishermen for the "inconveniences" caused by whaling operations in the area.[3] These groups also paid what was known as a *tatami-gin*, or "ground rent", for the use of their shore stations, look-out towers, and workers' accommodation huts (*naya*).

Money coming into the village in this way seems to have been spent in the same way as money earned from drift whales and catches of dolphins. In other words, it could be used for the benefit of the village as a whole, as was the case when earnings from the sale of dolphins were used for shrine and village entertainment. But the money could also, if the village so decided, be distributed more or less equally to all its households, as was the case when Hamasaki caught a school of dolphins (see Frame 4).

This system was partly retained after the Meiji Restoration of 1868, so that, with the development of the modern whaling industry, the traditional system of compensation payments continued. For example, Nata (in Fukuoka Prefecture) used profits from a drift whale in 1882 to construct an elementary school, which has since been known as the *kujira-gakkō*, or "whale school" (Suenobu et al. 1980:17). Again, when Tōyō Hogei decided to open a shore station in Ayukawa in 1906, it negotiated with Ayukawa village leaders, but in particular with the local Fishing Cooperative Association, that the company would pay an annual donation of ¥300 to the community.[4] This sum was used first to build a school in Ayukawa. At the same time, new forms of taxes were introduced by the authorities in order to support local administration. With taxes and compensation money paid by Tōyō Hogei and other whaling companies which established themselves in Ayukawa, the village was able to carry out a series of projects that included the building of roads, the provision of water to two hundred households, and the construction of a wharf on the water front. Not surprisingly, perhaps, we find that the annual income of the

---

[3] This rent was sometimes known as *ura-gin*—paid also by fishermen for the rental of sea space in another village's territory. In general, it seems that the fees paid to villages were fixed lump sums, but whaling groups (in the southwest of Japan) were also obliged to purchase a licence from the authorities, as well as pay a fixed lump sum tax (known as *unjō-gin*) for each whale caught. This tax varied according to the species of whale caught, its size (adult or juvenile), and the season in which it was caught (see Kalland 1988:213, 387, e.g.).

[4] That this sum should be a "donation" rather than a "fee" was very important to local villagers, since donations were tax-free.

village increased almost three fold between the years 1905 and 1907 (Oshika-chō 1988:160-168).

As has already been mentioned, with the establishment of local fishing cooperative associations (FCAs) following the introduction of the Fishery Law of 1901, the administration of the village's sea resources was transferred from the village itself to the cooperative, and in some cases compensation paid by whaling companies could thus be channelled through the FCA to its individual members. It has been up to the members of the cooperative to decide whether such money should be distributed to all members individually, as is often the case with compensation paid for pollution, land reclamations and so on (Befu 1980), or whether the money should be used to subsidize the FCA itself. This means that since the early part of this century, those immediately affected by whaling operations have also been the recipients of compensation. In Wadaura, for example, the whaling companies (Gaibō Hogei and Nittō Hogei) have rented shore facilities from the local fishing cooperative, and the money that they have paid has provided an important subsidy for the cooperative's activities. Similarly, in Ayukawa, all small type whales landed there have been sold through the local FCA's market and a commission of 3 per cent paid to the cooperative, so that its members have been similarly subsidized in that the fishermen pay lower commissions to the FCA than in most other fishing villages.[5] In Ukushima, where there in 1989 were plans to land pilot whales, the local FCA intended to charge an annual fee of ¥1 million for the use of a flensing station and ¥30,000 (or approximately 4.3 per cent of the total sales price) for each pilot whale sold through the cooperative.

The important point to note about the various ways in which fees have been paid is that the whaling groups or, more recently, companies were always obliged to pay some form of compensation for the inconveniences they caused to others. This was the case whether the owner or the operator was a member of the village or not. If he was an outsider, the payments of compensation became all the more important if he was to be admitted into village society. It is a well known feature of Japanese rural society that those wishing to settle in a village some time after it had been founded were often obliged to

---

[5] The importance of whale meat in the total revenue of the Ayukawa Fisheries Cooperative Association, and the subsequent undermining of the FCA's financial viability as a result of loss of sales, are discussed by Bestor (1989b:23-27).

wait several years or decades before they were given rights to irrigate land or fish at sea and, in some cases, such late-comers were never fully integrated into the local village structure, but remained only "half members" (see, for example, Shimpo 1976).[6] In this respect, the whaling companies were fortunate. By providing large amounts of cash in an economy in which cash was usually in short supply, they were able to buy their way into villages and carry out their activities to the mutual benefit of all concerned.

So far, we have looked at the external relations between local village and whaling company, and have shown the ways in which the former was able to benefit from the establishment of a shore station. Let us now turn to the internal organization of village life, in order to see how whaling actually affected the community as a whole.

Once a whaling company had established a shore station, it brought with it all kinds of other activities associated with whaling. Ships had to be fuelled, whale bones prepared into fertilizer at special plants, workers housed and fed. We thus find that, in the space of a few years, a village which had originally depended almost entirely on fishing became heavily involved in whaling. As we have already seen, much of the labour needed for specialized tasks was supplied from elsewhere by the whaling companies, but local inhabitants soon learned to participate in less skilled activities. In Ayukawa, for example, a number of wealthier households invested in the construction of sheds in which to boil whale bones for fertilizer, and they employed as workers members of their branch households and other relatives (thereby to some extent maintaining within the village the old pattern of household relationships based on seniority and wealth). Other households started up inns to accommodate seasonal employees from the whaling companies, while more and more people flooded in from elsewhere to take advantage of the new industry. In just over a decade, the total number of households in Ayukawa had tripled from just over 80 to about 250, and a number of companies (including a coal supplier and specialist in whale products) had established branch offices in the village. Given that the community also needed its fair share of entertainments to keep its thriving

---

[6] This was the case with Abe Gisuke from a nearby village who provided a large part of the capital required for the establishment of Ayukawa Hogei, but never became head of the Ayukawa FCA—a post at that time reserved exclusively for those born and brought up in Ayukawa (*tochi no mono*).

population happy in its spare time, in the space of little more than a decade following the establishment of the first shore station, Ayukawa had developed from a small fishing village into a town whose be-all and end-all was whaling. It had, in short, become a whaling community.

This economic integration of the village into the whaling industry was accompanied by various mechanisms—not strictly economic—which served to emphasize the people of Ayukawa's new-found identity and to create the concept of a local whaling culture in the town. One very important mechanism directly connected with the catching and processing of whales was the non-commercial distribution of whale meat among those immediately involved in the successful pursuit of a whale.

Non-Commercial Distribution of Whale Meat

It is well known that in the traditional village in Japanese rural society, relationships between main and branch households as well as between neighbours are continually cemented by the formal and informal exchange of gifts. We find that in the modern economy, too, gift giving forms an important means by which not only employees within a company, but whole companies themselves, are linked to one another. It is not surprising, therefore, to find that in fishing villages, a part of each catch used to be distributed by net or boat operators to their crews. Similarly, in the whaling industry, whale meat was frequently used—both as a form of payment in kind, and as a product which could then be exchanged among households with which those closely involved in the whaling industry had kinship and neighbourhood ties.

For example, whaling companies developed a system of meat distribution whereby all those on board a catcher boat were supplied with a fixed amount of whale meat for every whale spotted and killed. In LTCW and pelagic whaling, the distribution depended to some extent upon the position of each crew member, with gunners being paid twice as much as boat captains and chief engineers, who in turn received more than the rest of the ship's crew which was divided into officers (*shoku'in*) and ordinary seamen (*zoku'in*). Moreover, those working on catcher boats tended to receive more than those working on the factory vessel.[7] These amounts could be substantial, and one gunner in Arikawa recalled that during the heyday of whaling he received about 100-120 kg—mostly salted blubber—after one season in the Antarctic. In STCW, however, the distribution among crew members was equal—amounting to between one and two kilograms of high quality meat from every whale taken. At the same time, a distinction was maintained between those working on the catcher boats and those who flensed the whale on land, with the latter generally receiving less meat than the former. Indeed, in some cases, workers not usually employed at the shore stations, but taken on temporarily during very busy periods, were paid entirely in kind. The quality of the meat also differed, in that while those working on the catcher boats were paid in regularly shaped blocks of high quality meat, those working on the shore station tended to be given irregular or blemished pieces that would not fetch such high prices on the market (see Akimichi et al. 1988:42-43).

It was not only those directly involved in the hunting and processing of whales who received meat with the taking of every whale. Rather, the community as a whole also benefitted from gifts of meat. In the old days the whaling groups often paid part of their compensation to the "host" village with meat. For example, in 1730, a whaling group in Arikawa promised to give red meat "to the six coastal villages of Arikawa" three times during the season (Fujimoto et al. 1984:662), while, more recently, STCW owners in Ayukawa have frequently given whale meat to local institutions such as the Community Centre, the Old People's Club, local schools, the Children's Association, and the Fire Brigade, as well as to local temples and shrines (Akimichi et al. 1988:46-47). In short, whale meat has been an important focus of community identity in those villages in which whaling companies have operated and/or local inhabitants have been closely involved in whaling activities.

Another important example of this non-commercial distribution of whale meat among those living in a whaling community is to be found in the activities surrounding what is known as *hatsuryō*, or the first catch of the season, in LTCW and STCW. In Japan, we frequently come across people exchanging—say—*nashi* pears or *daikon* white

---

[7] Informants suggest that the distribution system (known as *buaikin*) was the same for whalers and fishermen, but that whaling has the added complexity of a special distribution rate for the flensing team.

radishes because they are both freshly in season and rare, and whale meat is no exception to this general rule of reciprocity. Thus, in 1838, for example, it is recorded that, after disposal of a right whale that had drifted into Hatsu-ura (Fukuoka), whalers distributed approximately three kilograms (5-6 *kin*) to three *ashigaru* soldiers stationed on the nearby island of Jinoshima (*Jinoshima-ura shōya kiroku*). Similarly, a whaling group at Fukuoka's Ōshima requested permission to give whale meat (*geiniku*) and one cask of tail meat (*hamo*) to the feudal lord through the magistrate for coastal affairs (Nishi Nihon Bunka Kyōkai 1988:213).[8]

Nowadays, too, an elaborate ritual of reciprocity surrounds the catching of the first whale in STCW. In the village of Ayukawa, for example, it is reported that during the several days between the start of the whaling season and the taking of the first whale, gifts known as "*omiki*" and consisting primarily of *sake*, are exchanged among three different sets of people involved in whaling: the vessel owner; vessels themselves and their crews; and individual crew members of whaling vessels (see Akimichi et al. 1988:43-51).[9] Local inhabitants give bottles of *sake* to those concerned because they wish to share in the distribution of whale meat following the first catch, and for every gift of *sake* received, the vessel owner is obliged to make a return gift of whale meat. This return gift is not limited to a single instance, but frequently takes place a number of times during the whaling season, so that in exchange for a bottle of *sake* someone may receive up to five gifts of whale meat, each weighing approximately one kilogram (Frame 7).

At the same time, gifts are also presented to the whaling vessels and their crews. Although exact records of these transactions are not available, it is clear that each boat may receive as many as fifty bottles of *sake* every year, and that these gifts are returned by members of the crew, who receive from their employer large pieces

---

[8] Fujimoto et al. (1984) provide a detailed list of gifts of whale products given to feudal rulers, religious institutions, workers, villagers, and so on.

[9] *Omiki* is a word that is used to refer to Japanese *sake* (rice wine) when used in a ritual context, in which the wine is first offered to the deities before being drunk by those participating in the ritual concerned. *Sake* itself has important ritual connotations in Japanese culture, especially as a means of purification, and the use of the term *omiki* here indicates the importance that those living in a whaling community attach to this particular form of gift giving.

Frame 7: Gift-Giving in Ayukawa

The distribution of gifts in Ayukawa during the period immediately preceding the first STCW catch of the year in mid-April has already been documented (see Akimichi et al. 1988:43-51), but it is useful to review that information here. In April 1986, one of the boat owners received a total of 156 bottles of *sake*, together with a few bottles of whiskey and crates of beer and Coca Cola. In return, he distributed approximately 1 kg packs of whale meat in the following manner:

Table: First Catch Gift Giving, 1986

Received	*Bottles of* sake			*Whiskey*		*Beer*	*Coke*
	1	2	3	1	2	(12)	(20)
Returned							
1x	-	5	-	-	-	-	-
2x	1	37	-	3	-	1	1
3x	2	31	1*	-	-	2	-
4x	-	1	-	-	1	-	-
5x	-	1	-	-	-	-	-

Note: * indicates that this person also gave one crate of beer

It can be seen that the boat owner may return the original gift on as many as five separate occasions, and usually does so between two and three times. In other words, the 86 people who participated in the first catch gift giving received approximately 200 kg of whale meat in return. Since the retail price of first grade *sake* at the time was just under ¥2,000 a bottle, with most people giving two bottles each, and given that the retail price of high grade whale meat was between ¥2,500 and ¥3,000 per kilo, it can be seen that the return gifts were between 1.5 and 2 times higher in value than the original presents.

What is of particular interest is the way in which return gifts were distributed, depending on the relation of the gift giver to the boat owner. The owner himself divided all those concerned into three categories—relatives, friends (acquaintances and neighbours), and business associates—and distributed return gifts accordingly.

The interesting anthropological point here is that the combined number of relatives, friends, acquaintances and neighbours came to more than twice that of business associates with whom gifts were exchanged. Moreover, it is apparent that, in the networks of exchange set up in Ayukawa through this non-commercial distribution of whale meat, relatives were valued more highly than friends and neighbours, who themselves were placed on a slightly higher level of reciprocity than were business associates. In March 1987, when the owner launched a new boat he received gifts of *sake* from 134 people and—regardless of the quantity received—he gave 2 kg of whale meat on two occasions to everyone except relatives—to whom he gave meat every time he caught whales that season. The data suggest that balanced reciprocity has been customary with business associates, while a more generalized reciprocity (Sahlins 1972) has been practiced towards friends, neighbours, and in particular relatives.

of meat weighing 30 or 40 kilograms which they then redistribute to those concerned. It is known that upon receiving gifts of whale meat, local inhabitants then used to redistribute these gifts in smaller quantities to relatives, neighbours and friends, so that at the beginning of every whaling season, communities such as Ayukawa were involved in one vast cycle of gift exchange that appears to have rivalled even the famous *kula* ring discovered among the Trobriand islanders (Malinowski 1922). Hardly surprisingly, therefore, local inhabitants in whaling communities have commented that whale meat is not for buying, but for exchanging with others. In other words, the distribution of whale meat in a whaling community such as Ayukawa is closely connected to long-term ongoing ties between individuals and households therein. To "eat" whale meat thus becomes an important metaphor for social exchange, and for community identity as a whole.

Whale as Food

Our analysis of the non-commercial distribution of whale meat leads to a discussion of the role played by whale in Japanese local cuisine. Whale meat has been particularly prized by the Japanese for many centuries, receiving "honourable mention" in a famous cook book of the Muromachi Period published in 1487. In 1832 a special whale cookery book, *Geiniku chōmihō*, was published in Kyūshū, and this divided the whale into seventy named parts, each with detailed information about methods of cooking and nutritive value. Whale meat—including blubber, skin, cartilage, fluke, intestines, and genitals—has for centuries been used as food. During the first post-war years whale meat accounted for 47 per cent of the animal protein intake by Japanese, many of whom are convinced that whale saved them from a major famine. Indeed, some of their attachment to whales and whale meat possibly stems from this belief.

The importance of whale meat has given rise to a rich culinary tradition. The quality of the meat is finally graded, and various parts are regarded as suitable for different dishes. Regional food preferences have emerged as a result of the history of whaling in particular communities. Such preferences exist not only as a result of the particular species traditionally caught, but also in terms of method of cooking. In Arikawa, for example, the most cherished whale meat in the past was that from right whale. Because this meat is no longer available, salted blubber of fin whale has become a new favourite. This was the meat many of the local whalers brought home at the end of each season's whaling in the Antarctic.

In Taiji, on the other hand, people have developed a special liking for pilot whale, which is often eaten raw as *sashimi*. In Wadaura a local speciality consists of dried, marinated slices of Baird's beaked whale (*tare*), but people in Arikawa find this meat "offensive" because of its allegedly "strong smell". Baird's beaked whales are not eaten in Abashiri or Ayukawa either, and such whales landed there used to be sent to the Wadaura area for processing and consumption until the moratorium disrupted the marketing structure (cf. Chapter 9). The inhabitants of Abashiri and Ayukawa prefer raw red minke whale meat. A special New Year's dish in Abashiri is soup boiled from salted blubber. The meat of sperm whale is preferred in some areas of northeastern Japan, while Arikawa people find sperm whale meat fit only to be eaten dried or in fish paste (*kamaboko*).

Local taste varies, too, with other cetaceans. Dolphins, for example, are extensively eaten in some communities in the southwest, particularly in Taiji and in Arikawa, where it is either salted and dried, made into pickled meat, or boiled and used in *sukiyaki*. However, there is little demand for this kind of meat in many other parts of Japan, and marketing has been a serious problem for those attempting to develop dolphin hunting in Hokkaidō. One of the reasons for some people's poor regard for this kind of meat is a lack of proper knowledge about how best to prepare the meat. But this knowledge is gradually spreading. Such diffusion of cuisine is sometimes deliberate, as when people in Iki recently learned to cook dolphin from specialists in Arikawa.[10] Thus, with the reduced scale of whaling there are indications that whale cuisine will become more uniform throughout Japan.

Few things are as symbolically laden as food, and local cuisine is one of the strongest markers of social identity in Japan. For example, people in Nagasaki are proud of their *champon* noodles, those in Hamamatsu of their eels, those in Akita of their *kiritanpo* barbecued riceballs, and the inhabitants of Uji in Kyōto Prefecture of their green tea. People in the two main urban centres of Tōkyō and Osaka will also use food preferences (among other things) to help establish separate identities.[11] Food often figures as the "special products" of localities, and the alert foreign traveller will notice that many railway stations in Japan sell lunch-boxes (known as *eki-ben*) featuring local specialities. Food is thus an important element in building a positive image of local communities, or what are nostalgically referred to by the Japanese as *"furusato"* (or "hometowns").

The various ways of preparing whale meat have become important means by which whaling communities express their identity, and might in fact be one reason why whalers are so reluctant to eat whale dishes from communities other than their own.[12] Whalers from different parts of Japan never seem to grow tired of discussing local whale cuisine. When housewives' associations on Nakadōri Island (in

---

[10] Iki Island had long been annoyed by large schools of dolphins, which interfered with its fisheries. Some years ago, therefore, the villagers got financial assistance from Nagasaki Prefecture and invited people from Wakayama Prefecture to come and teach them how to drive dolphins into bays. The animals were originally meant to be sold to a fertilizer plant. In order to make better use of this resource, however, people from Iki travelled to Arikawa where they studied dolphin cooking, and the owner of a whale restaurant in Arikawa spent two weeks in Iki teaching his skills.

[11] In Tōkyō, food is said to be "dark" and "thick", while in Ōsaka it is "thin", depending on the amount of soy sauce used. Tōkyō people regard Ōsaka food as "sweet" because of the amount of sugar added. Ōsaka people thinks that Tōkyō noodles are inedible—which is what Tokyoites think of Ōsaka *sushi*!

[12] There are reasons for believing that the reluctance of some people to eat dolphin meat is based more on psychology than on actual taste preferences. It is not part of their culture to eat dolphins. Indeed, dolphin meat might find a more ready market if it were sold as "whale" (*kujira*) rather than "dolphin" (*iruka*). This testifies to the importance of food for people's identity and to the difficulty of substituting whale meat with other kinds of food.

the Gotō Archipelago) meet in order to prepare and display their specialities, the Arikawa association always takes care to present whale dishes.[13] People also regard it as very important to be able to serve whale meat dishes to tourists in order to promote a community's image of being a "whaling town". When Ayukawa was allowed to buy some of the minke meat obtained during research whaling in the Antarctic in 1988, a substantial part of it was channeled to hotels, inns, restaurants, and other institutions catering to tourists, including the famous shrine on the offshore island of Kinkazan, which received a share so that it could continue to serve whale dishes to pilgrims there.

When whale meat was cheap, it was consumed almost daily. With the present dearth of supplies, however, prices have soared and many people add only a small amount of meat at a time, "just to get the flavour of it". Whale meat has thus become almost a kind of seasoning, and most people save what little they have left for special occasions. In whaling communities whale meat is an indispensable part of any type of community gathering or celebration. It is extensively served at important *rites de passage*—such as a child's first day at school, a wedding, funeral, or memorial service for the ancestors—as well as to celebrate the building of a new house, and so on.

Whale meat is also a typical food for the New Year and other important annual events. As we saw in Chapter 2, about a fifth of the annual sale of whale products in Arikawa occurs during the New Year season. August, which is also marked by high sales of whale meat, is the month of *obon*, or "All Souls' Festival", when the ancestors are believed to return to the houses in which they once lived. This is the season when people try to visit their places of birth (*furusato*) in order to meet both relatives and ancestors. A third peak in Arikawa's whale meat consumption (which is becoming less and less apparent, however) occurs in late March or early April, and reflects an old, but dying, tradition in which Girl's Day (*osekku*) is celebrated in combination with cherry blossom-viewing (*hanami*) when special

---

[13] In 1988, the following dishes were presented: (1) *Namasu*, a dish made of *wakame* (a seaweed) and blubber (*shiro-kujira*) over which hot water had been poured with sugar and vinegar being added; (2) *kujira-gohan*, rice boiled with salted blubber and burdock root (*gobō*); (3) fried whale meat cutlet (*geiniku no katsu*), a dish that often used to be eaten at school lunches.

three-layered lunch-boxes (*jūbako*) are prepared. In the past, on passing her or his first *osekku*, a child would be given such a lunch-box to keep, often by a relative other than the immediate family (Kalland 1989:110-113).

Whale meat is thus consumed by the villagers in connection with most religious events. Not only this, but meat is also frequently donated to the deities to "eat". In STCW, and to a lesser extent in pelagic whaling and in LTCW, it has been customary to donate a part of the tail to the "god shelf" (*kamidana*) on the boat. Many whalers do the same to their *kamidana* at home. Some people will even give whale meat to the altar for the ancestors (*butsudan*).[14] The meat given to deities and ancestors is later eaten, of course, in a communal meal—in much the same way as Christians partake of Holy Communion.

Food then, affects the whaling community in a number of ways. Whale meat gives the community an image, which is now needed in order to promote tourism, for example. Furthermore it gives the villagers an identity—an identity different both from those of other villages in the vicinity and from those of other whaling villages. Whale meat also helps the villagers to mark time, by being used to emphasize all important events in the life cycle as well as in the annual cycle. Finally, whale meat is used in communication, both among members of each community (as we have seen with the non-commercial distribution of whale meat as gifts), and with the deities and ancestors. This brings us to the importance of whaling to rituals and beliefs in these whaling communities.

---

[14] Those studying Japanese society and culture may find this odd, since usually only vegetarian food is given to the ancestors. However, extensive use of whale meat as offerings to deities and ancestors has been reported from several whaling communities (Iwasaki 1987; Akimichi et al. 1988), although this particular use of the meat seems to be limited in Arikawa. The Shintō priest there sometimes offers salted whale meat to his household Shintō altar, regarding whale as a "fish", but he never uses whale meat as offerings (*sonaemono*) during festivals. Some informants explicitly said that they did not offer whale meat either to the *kamidana* or to the *butsudan*. Whalers and their wives, however, used whale meat as offerings to shrines during their pilgrimages, and this meat was later eaten during their communal meals.

Rituals and Beliefs

One of the central institutions in most Japanese local communities is the Shintō shrine which, as we mentioned earlier, is dedicated to the tutelary deity (*ujigami*), and which to a great extent defines the local community. Both the deities enshrined and the ritual practices involved are unique to each shrine. There are thus many differences in the ways in which communities and households perform their whaling-related rituals, but there are also common themes therein. For instance, it is essential to be on good terms with the Shintō deities, one's ancestors and the souls of deceased whales in order to make a safe and financially successful journey. The ties among the whalers themselves, among their wives, between whalers and their families, and between man and whale have been strengthened through a number of rituals (see also next chapter).

Frame 8: A Whale of a Tale!

According to the *Ukuchō Kyōdo-shi* (pp. 418-420), there is a legend for all islands in the Gotō archipelago known as the "Monkurō Whale". According to this tale, Yamada Monkurō, third generation owner of the Yamada net whaling group, was very successful, but in the winter of 1715 his group failed to catch a single whale and his workers began saying how good it would be to catch one before the end of the year. The New Year came and went, and they were well into the first month when, on the 21st, Monkurō had a strange dream in which a whale appeared with her calf. "I am on my way to Daihōji Temple in Gotō to pray with my child," she explained (Daihōji [*lit*: "Great Treasure Temple"] being an extremely famous temple belonging to the Shingon Sect. It was said that if one were to present a child's umbilical cord to the temple, one would receive "great treasure"). "Please don't catch me until we have completed our pilgrimage."

Monkurō awoke from his dream and thought: "I see, today is a festival day. If a whale really does pass by, I must take care not to catch it." Eventually day broke and he told his comrades about his dream. "Today we may very well sight a whale," he said. "But if it is with calf, then don't take it—just this time." The whalers, however, were unimpressed by what Monkurō asked of them, since they realized that it was just a dream, and they therefore contented themselves with non-committal replies.

After lunch they were all having a rest when the look-out started shouting excitedly. Everyone got up. It was the whale for which they had been waiting for so long. "It's a whale with calf," somebody said. "Come on! Let's get it!" They cried with great bravado, already forgetting what they had just been asked by their boss. So they donned their *hachimaki* head-bands to make themselves look really manly, and jumped up to launch their *seko-bune* chase boats and *sōkai-sen* net boats into the open sea.

They followed the whale—a blue whale, 33 *hiro* (about 60 metres [*sic!*]) in length. "Don't let it escape!" The whalers all cried. The nets were lain—one, two, three. The big harpoon was thrown at the whale, which thrashed around trying to escape. But the whalers stuck to it, failing to notice that the skies in the west were suddenly turning ink black. The battle between man and whale continued. The black clouds spread all over the sky and the wind rose. The rain, even sleet, came down as the weather changed. Ukushima was already far away and very small on the horizon. In its struggle with death, the whale finally shook itself free of the nets, broke off the harpoon, and escaped. The waves began to foam white, rearing up like mountains one after the other, trying to swamp the boats. The captain called to the men to turn back for home and they all started rowing for shore. But the boats got separated from one another and nobody knows what happened to them.

The night passed with people on the island in great distress. The next day dawned beautiful and clear, with the wind quite calm, and it was found that 72 people had been carried away by the waves and lost at sea. It was the greatest disaster since the start of net whaling. It is said that the Yamada crest was imprinted on the whale's back and that Monkurō never established another whaling group again.

It is hard to tell just how true this tale is but there is still today a memorial in the cemetery at Taira on Ukushima island, commemorating the deaths of these 72 whalers almost three centuries ago. There are many similar stories about malevolent spirits of whales, and some of them are known throughout Japan, with minor local variations. In Taiji, for example, a legend recounts how 111 whalers lost their lives in 1878 in an attempt to catch a right whale with a calf. As we have seen, this accident has become part of the community's cultural heritage, and is used to give Taiji its peculiar identity. At the same time, the wrath of the deities evidenced when the whalers broke the taboo against catching a whale with calf has also served to reinforce the validity of the taboo—a taboo which may well function primarily as a means of preventing overfishing and of conserving whale stocks.

According to the prevalent world view in Japan all animals are endowed with souls so that, in this respect, there is no fundamental difference between animals and human beings. Moreover, men can become indebted to animals, as well as to deities and fellow human beings, and whalers become indebted to whales "who have given their lives so that we can live". The whales are also gifts from nature, which itself is believed to be infused by Shintō deities (*kami*). Thus, whaling activities become intimately bound up with religious beliefs and, as gifts from the deities, whales have to be fully utilized, for to do otherwise would be an insult to both deities and whales. To repay the whales for sacrificing their lives, whalers have furthermore to take care of their souls, or else these whale souls can turn into "hungry ghosts" (*gaki*) which might cause illness, accidents or other misfortune.

It has, therefore, been the practice in many whaling communities to treat the souls of whales in the same manner as the souls of deceased human beings. The whales have been given posthumous names (*kaimyō*), which have been inscribed on wooden memorial tablets (*ihai*) and included in death registers (*kakochō*). Tombs and memorial stones can be found in at least 48 places, from Hokkaidō in the north to Kyūshū in the south, and at least 25 festivals (*matsuri*) and memorial rites (*kuyō*) are held annually in honour of whales (Akimichi et al. 1988:55-56). One tomb at Kōganji in Nagato (Yamaguchi Prefecture) has been designated a national historical monument. Built in 1962, it marks the burial place for 75 foetuses found in the wombs of whales caught before 1868. The temple of Kōganji is in fact dedicated to whales and, in one of the most elaborate memorial services known, Buddhist priests recite sutras for several days each April in order to help the whales' souls be reborn into a higher level of existence. These rituals give the local residents both the feeling of a common heritage and meaning to their lives.

The main community festivals, which are focused on the village shrine, aim at securing abundant catches (*dai-ryō*) and safe voyages for the whalers. Although there are no particular Shintō deities for whaling, a number of deities which are enshrined therein are made relevant also to whaling. For example, when whaling came to Ayukawa, the deities of the local Kumano Shrine and Inari Shrine, which until then had served the fishermen well, also came to be used to protect whaling and whalers. In other communities new shrines and deities have been introduced in addition to the old ones. This is the case in Arikawa, for example, where the main whaling festival has been held for the sea goddess Benzaiten, who was brought to the community from Kamakura in 1689. Whether the shrines focussed on during the main festivals are new or old, these festivals are seen as crucial for the welfare of the community. Through the annual festival the ties of mutual dependence and reciprocity are renewed, and the "contract" between man and deity are "signed" by a communal meal.[15]

At the household level there are a number of ritual observances, some of which are performed daily. Most whalers have a Shintō "god shelf" in their house, where a short morning ceremony is performed and to which gifts of rice, water and tea are offered. As already mentioned, whale meat can also be donated when whales have been caught, both to express the whalers' gratitude towards the deities, whales and nature in general, and to request further success in the future.

When their husbands are out whaling, housewives tend to carry out these rituals with greater fervour.[16] As a result, morning rituals tend to be longer than in the off-season. The wives of whalers also frequently go to other shrines together in order to make offerings and pray for good catches and safety. In Ayukawa, several wives meet every morning, as long as their husbands are active in STCW, in order to go to the local Inari Shrine, while in Arikawa the wives of whalers engaged in LTCW and pelagic whaling used to go on a pilgrimage to the local Kompira Shrine on the 10th of every month as long as their husbands were away.[17]

Gratitude and requests for success thus characterize these Shintō

---

[15] Both successes and failures are explained in relation to the divine. Failure to fulfil one's part of the deal can cause withdrawal of divine support, thus leading to poor catches or accidents. On the other hand, certain deities may also fail to live up to expectations, and so lose human support. Festivals are therefore part of an on-going process of communication between man and deity.

[16] Rituals by wives are not limited to whaling, but can be found in other more or less dangerous occupations. For an explanation of misfortunes connected with women's failure to perform such rituals on behalf of their husbands, see Moeran (1985:84, 227).

[17] Kompira is a deity for fishermen and seafarers in general. There is a famous Kompira Shrine on Shikoku, and most Japanese ships, including whaling boats and factory ships, have amulets (*ofuda*) from this shrine on their "god shelves".

rituals connected with whaling. Although these are also important elements in Buddhist rituals, the latter contain other elements as well, for they are basically memorial services (*kuyō*). During a typical memorial service the temple priest recites a special sutra (*osegaki-kyō*) read also for those who have lost their lives at sea. Such services have a number of implications, and people may have different interpretations of the rites. The temple priest in Arikawa, for example, performs the memorial service in the belief that the whale will be released from rebirth in this world and enter Paradise as a "Buddha" (*hotoke*), although he supposed that some people might believe that the whale would be reborn as a human being. This rite also serves to secure safe voyages in the future by preventing whale souls from becoming hungry ghosts. Finally, memorial services are held to ensure that the whalers, and the gunners in particular, are forgiven for the sin involved in taking life. Not surprisingly, the memorial services carry special meanings for the gunners, and many pelagic gunners went directly to their local temple when they returned home, in order to conduct memorial services for the whales they had killed. The whales are also usually included in memorial services held for the dead during the equinoxes in special services for those who have died during the previous year (*ojūya*), and during ancestral *obon*.

Some temples, such as Kōganji in Nagato and Kannonji in Ayukawa, have special wooden memorial tablets (*ihai*) for whales, whereas in other places the souls of whales may be included in tablets made for wandering spirits in general. The temple priest may perform services for these wandering spirits at the same time as he does so for the household's ancestors. The Buddhist priest in Arikawa is also called on to perform services on the beach for wandering spirits in connection with poor catches of fish, and it appears that such services were probably also conducted in the old days of net-whaling. There is, moreover, a grave for wandering spirits at Tateishi near Arikawa, a place where in the past whale embryos were buried with great care. On such occasions the priest would be called immediately and a funeral resembling that for human burial would be performed with all the harpooners dressed in formal clothes.

Memorials are frequently carried out at temples, but many whaling communities have special memorial stones or monuments (*kuyōtō*), where such services are performed. Some of these date back to the seventeenth and eighteenth century—such as the one in Arikawa where ceremonies are no longer performed—while others are newer, such as the whale monument in Taiji, which is the focal point for the movement to protect whaling there. Memorials are also performed in front of the ancestral altars at whalers' homes. Practice varies from household to household, but it is not uncommon to find whalers offering daily prayers to the killed whales. To some whalers such prayers are expressions of gratitude rather than intended to secure the whale a rebirth in Paradise. As one whaler in Abashiri said: "If the whales had souls, how could we kill them?" To this particular man there were fundamental differences between ancestors and whales. To most people, however, whales have souls and these souls have to be comforted at the ancestral altar together with one's own ancestors.

Rituals also serve to give each community its distinct character. The set of Shintō deities worshipped is unique to a community and the festivals differ as well. However, although there may be local differences in the deities worshipped and in the timing and performance of rituals, they are all variations of common themes, based on a conception of the whale as a creature with an immortal soul, and on a world view stressing the interdependence of supernatural, human, and animal worlds.

Conclusion

In this chapter, we have focussed on those social and cultural mechanisms that have helped give rise to what we have termed a local integrated whaling community. Our starting point has been that the Japanese household in general is an economic unit comprising a number of individuals, related by blood, marriage and/or adoption, and that the Japanese village has always been an administrative unit whose main function has been to look after common property and resources. In this respect, neither the whaling household nor the whaling community has deviated from the norm. Whalers, their wives, children and parents, all live together in residential units which are defined by their economic status; whalers' households together form a community in which the administration of those employed in the hunting and processing of whales takes priority.

It is characteristic of man, however, that he does not simply live and work with other people. Rather, he creates a set of ritual beliefs and social and cultural practices which not only give meaning to his otherwise "economic" existence, but give rise to an ideology of "community", by which his co-residence with his fellow men is justified and explained. Thus we have found in our discussion of local whaling culture that there are certain mechanisms which whalers use to establish their community identity. By distributing the meat of the whale both to those actually involved in its hunting and processing and to relatives and neighbours living in the community, whalers are able to strengthen relations built up in their place of work, on the one hand, and of residence, on the other. In short, they are able to combine the economic and residential functions of the village through presents of whale meat, which serve to link almost all households to one another and hence establish a common identity focussed on the object of their livelihood, the whale.

We have found that this sense of local identity is supported in other ways as well. Just as rural farmers are convinced that the main product of their activity, rice, is a vital aspect of their diet, so, too, do we find that whale meat occupies a vital role in each whaling community's cuisine. Whale meat (or the lack of it) and the ways in which it can be cooked and prepared are the source of conversations and of community pride. "Others may do things their way, but we do it this" is the kind of attitude usually displayed by people living in different communities. Thus, although there is a plurality of local culinary tastes and attitudes, whalers living in different communities use their food as a means of asserting their community identity and hence independence from other whaling villages, of whose existence they are, of course, very much aware. Farmers may believe in the inherent value of rice, and young city people in salvation through McDonald's hamburgers, but whalers have whale meat. That is what sets them apart.

This sense of local identity is also strengthened and upheld by rituals and beliefs surrounding the hunting and processing of whales. Here the general practices of both Shintō and Buddhism are brought into play, and we have seen how the whale has been incorporated into the Shintō pantheon, on the one hand, while being memorialized according to Buddhist beliefs, on the other. In this respect, whaling communities adapt to general practices common throughout Japan. At the same time, by establishing separate deities for whales in different villages, whalers are able once again to assert their independent local community identities. This independence is, of course, further emphasized by the fact that it is the whale rather than anything else, which is the focus of their beliefs.

In this chapter, therefore, we have argued strongly for the notion of a local whaling culture—which facilitates transfers of knowledge across various forms of whaling as well as between sets of activities. At the same time, however, much of the evidence adduced has suggested that local cultures tend to be plural and that the whaling culture in Arikawa, for example, differs from that found in Taiji, which in itself is not the same as that found in Ayukawa or Abashiri. The question then arises: if local whaling cultures are plural by nature, what is it that gives rise to an integrated whaling culture in the singular? It is to this question that we will now turn.

CHAPTER 8

Whaling Culture & Whaling Companies

One of the main institutions linking the two sets of whaling activities (i.e. hunting and processing) as well as the three forms of whaling (i.e. pelagic, LTCW, and STCW) is that of the whaling company. This linking was achieved on several levels. Firstly, the various whaling companies have been linked with one other through financial arrangements, at times operating jointly on whaling grounds, and all participating in the same industrial associations. Secondly, a large number of people from various places were brought together as a result of the systems of recruitment employed. Employees were firmly integrated into the company through such organizational strategies as rituals, songs, company newspapers, and former employees' associations, and together these created the kind of in-group feeling that is so often a feature of institutions (and, of course, of companies in general) in Japanese society. Finally, the company has helped transfer knowledge of whaling both spatially and temporally—from one region to another, and from one generation to the next. In this respect the whaling company has, of course, been central in the training of whalers.

Linkages Between Whaling Companies

The overall structure of the Japanese whaling industry is complicated and beyond the scope of this book, but certain points need to be made about it if we are to understand the way in which whaling companies have contributed towards an integrated whaling culture in Japan. In the first place, all types of active whaling discussed here (pelagic, LTCW and STCW) have been carried out by whaling companies of one kind or another. These include not only the big corporations involved in large scale commercial whaling in the Antarctic, but also those small family run enterprises engaged in STCW. Thus, at one extreme, we find companies like Nissui and Taiyō employing several thousand whalers each during the heyday of whaling (together with several thousand more in their fishery industries), while at the other, there are STCW enterprises with fewer than 20 employees each.

Related to this is our second point that these two types of company are very different in both their history and their organization—a matter to which we shall return when we discuss company culture later on in this chapter. However, we should note here that there has been a considerable number of mergers and acquisitions of whaling companies by other whaling and/or fishery companies during the past century. We have already mentioned the existence of Ayukawa's own LTCW whaling company, Ayukawa Hogei, which was then taken over by Sumatra Takushoku, before merging with Kyokuyō Hogei five months later. Similarly, Tōyō Hogei was originally formed by four main whaling companies, which then absorbed five smaller companies, before being merged with Nihon Sangyō to form Nihon Hogei in 1934. Two years later, this latter whaling company further merged with Kyōdō Gyogyō, which changed its name to Nihon Suisan (or Nissui) in the following year (see also Chapter 4).

Thirdly, a number of whaling companies have over the years conducted joint operations, both in the Antarctic and the North Pacific. The aims of such joint ventures have been various— involving the acquisition of skills, providing capital, spreading financial risk, and adjusting to industrial regulations and licensing policies.

Such operational linkages are, in fact, historically quite old for joint ventures were common during the days of net whaling. Initially they were a strategy used to acquire new skills, as—for example— when entrepreneurs from Taiji and other villages along the Kumano Coast were invited by feudal lords and local entrepreneurs in Kyūshū to start whaling there. Later, financial considerations came to be the dominating motive. The huge capital investments required to organize net groups at this time tended to some extent to destabilize operations, and many groups went bankrupt even before they had successfully hunted their first whale. Those net groups which survived were often heavily indebted both to the feudal authorities and to large merchants in Ōsaka and elsewhere. Joint ventures could resolve this problem to some extent in that risk was spread among several entrepreneurs. Frame 9 shows how operators of net whaling groups during the Tokugawa Period in fact found themselves entering into the same kinds of joint venture relationships that later characterised the mergers and acquisitions of 20th century whaling companies.

Frame 9: Arikawa Net Whaling Groups

Joint ventures seem to have been particularly common in Kyūshū (Hidemura 1952a:69). In Arikawa, for example, where it was at times difficult to find people willing or able to run a net whaling group, and where the authorities had to intervene and take over the operations themselves, the Eguchi household cooperated with a number of entrepreneurs from other villages.

Eguchi's first group, which used the harpoon method, was formed jointly with an entrepreneur from the Kumano Coast in 1626, clearly in an effort to gain expertise. This might have been the reason for his descendant's collaborating with Yamada Mobei from Ukushima between 1691 and 1693. But thereafter it was clearly financial difficulties which obliged the Eguchis to invite Nakao Jinroku from Yobuko (Saga Prefecture), and—later—Shibata Rokuzaemon and Hirozaemon from Uonome (facing Arikawa) to join their group.

Period	Operator(s)
1684-1691	Yamada Mobei (from Ukushima)
1691-1727	Eguchi (1691-1693 with Yamada Mobei)
1728-1737	unknown
1737-1745	Eguchi & Nakao Jinroku II (from Yobuko)
1745-1751	unknown
1751-1752	Defunct and under re-organization
1753-1757	Eguchi & Nakao Jinroku III
1757-1765	Nakao Jinroku III
1765--->	unknown
---> 1799	Yugawa Genjiuemon
1800 --->	Ukushima Genuemon (from Yobuko)
---> 1815	unknown
1815-1817	Eguchi
1818-1827	The feudal authorities
1827-1836	unknown
1837-1841	The feudal authrorities
-->1843-->	Eguchi & Shibata Rokuzaemon and Hirozaemon (from Uonome)
1846-1866	unknown

Sources: Hidemura (1952a); Nakayama (1987); Yoshida (1972)

In the present century, similar kinds of financial security and the acquisition of skills have been sought through mergers and joint ventures. Financial considerations have been particularly important in company mergers during the first decade of this century when the industry was consolidated, and again in the 1930s when huge investments were required in order to send fleets to the Antarctic. At the same time, it seems that Kyokuyō Hogei bought up Sumatra/Ayukawa whaling company in order to acquire skilled whalers—in particular, gunners—whom they could then send on their first Antarctic expedition.

After World War II, joint ventures were used to reduce risk when Japanese whaling companies have established whaling bases abroad, in such places as Canada and Brazil (cf. Chapter 4),[1] but joint ventures in this century are first and foremost a response to restrictions imposed on the industry by Japanese government authorities, who have been rather strict in issuing licences for whaling, particularly in waters close to Japan. Thus, before World War II, the authorities limited the number of pelagic fleets in the North Pacific to just one (run by Nissui), and the three large companies were compelled to send a joint fleet there in 1940 and 1941.

This restrictive policy continued after the war. From 1952, when the first fleet was sent to the North Pacific, until 1962 all the large whaling companies operated joint fleets, with only one being sent in 1952 and 1953. Kyokuyō, which was not given a licence to whale for baleen whales in the Antarctic until 1956, provided the factory ship for the fleet, with the four catcher boats coming from three companies. Between 1954 and 1961 two fleets operated jointly (Table 3).

From 1962, each of the three main companies operated their own fleets. Nittō Hogei provided catcher boats and the factory ship *Nittō-maru* for the Nissui fleet; Nihon Kinkai catcher boats for the Taiyō fleet; and Hokuyō Hogei catcher boats for the Kyokuyō fleet.[2] Then,

---

[1] We might note that, when establishing overseas subsidiaries, Japanese whaling companies frequently practised a form of personnel redeployment characteristic of Japanese company organization as a whole, whereby employees who had reached retirement age were placed in senior managerial positions in the subsidiaries abroad.

[2] According to the international convention, a fleet which had hunted in the Antarctic could not catch baleen whales on other whaling grounds before the elapse of one year. This restricted the Japanese fleets in the North Pacific considerably. The Kyokuyō fleet, which until 1956 did not go to the Antarctic, was licenced by the

Table 3: The Structure of the North Pacific Fleets, 1954-1961

Factory ship operated by	Catcher boats operated by	Quotas given by Japanese authorities
1. Kyokuyō	Taiyō (3), Nissui (3), Kyokuyō (1), Nittō (1), Nihon Kinkai (1)	800 BWU + 200 sperm
2. Taiyō/Nissui (alternate years)	one each from Taiyō, Nissui, Kyokuyō, Nittō, and Nihon Hogei	0 BWU 1300-1500 sperm

Source: Adapted from le Grand et al. (n.d., pp.20-21)

in 1976 the number of fleets was again reduced to one when the Japanese Government had to accept the reduced quotas set by the IWC. This was the year in which Kyōdō Hogei was established, with Nissui, Taiyō and Kyokuyō each owning 28.2 per cent of its shares, Nihon Hogei and Nittō Hogei 1.5 per cent each and Hokuyō Hogei 1 per cent. The remaining 11.5 per cent was bought up by financial institutions (le Grand et al., n.d., p.69). From that year until pelagic whaling came to an end in 1987, this type of whaling was conducted jointly by all companies involved through Kyōdō Hogei, and there was a complicated system by which the factory ship was paid for flensing whales, whose meat was then "sold" at an agreed price back to participating companies which had provided the fleet with catcher boats and caught the whales in the first place (Tatō 1985:92-103).

This kind of joint venture has not been limited to large scale operations in the North Pacific, but has also involved STCW vessels. Although each of the boats has its home port in Taiji, Wadaura, Ayukawa, or Abashiri respectively, they are free to hunt whales wherever they wish, provided that they adhere to the limitations imposed on them regarding seasons and quotas. Thus, all nine boats

---

Japanese authorities to catch baleen whales in the North Pacific, whereas the Nissui/Taiyō fleet was only allowed to catch sperm whales. From 1962, the three big companies were granted licences to catch both baleen and sperm whales as long as they kept to the rules of the Convention. Consequently, Taiyō and Kyokuyō made an agreement to the effect that Kyokuyō would take both companies' quotas of baleen whales, while Taiyō took both quotas of sperm whales. This arrangement lasted from 1965 until 1969 (Tatō 1985:191).

operated in Hokkaidō waters during the summer of 1987, and all but the Abashiri boats hunted off Ayukawa during the spring of the same year (Akimichi et al. 1988:21-23). This means that crews were able to observe one another's behaviour and exchange information about whale sightings (thereby economizing the overall hunting effort) during these seasons together. Since 1988, when the moratorium was extended to cover the minke whales hunted by small type coastal whalers, this kind of cooperation has taken on a new dimension, for several STCW companies have found themselves obliged to work in pairs (in other words, two companies operating one boat), thereby reducing the number of operative STCW boats (and whalers) by half.

Creating a Company Culture

Having briefly looked at the overall structure of the whaling industry, let us now turn to the whaling companies themselves. The Japanese company has often been described as an encompassing organization which acts to enclose its own employees in an identifiable group (cf. Nakane 1970; Rohlen 1974; Clark 1979; and others). It is more than a place of work where one receives one's paycheck in return for labour given. It is a social universe, a place where an employee finds his or her social identity, makes friends and seeks comfort. The paternalistic qualities said to exist in Japanese companies have received wide publicity (both at home and abroad), and the majority of Japanese workers have been found to prefer working in companies with close and emotional ties between employees and employers (Marsh and Mannari 1976:317). To foster such sentiments, which are so often a feature of Japanese society in general, companies make use of certain strategies such as rituals, songs, company newspapers, and former employees' (or "Old Boys'") associations (*OB kai*). Whaling companies are no exception to this general pattern.

The Paternalistic Whaling Company

It is difficult to generalize about Japanese companies, which are often as different from one other as from companies in other parts of the world. This is equally true of whaling companies. For one thing, as we have already pointed out, they have always differed considerably in size—a fact naturally influencing the extent to which the companies concerned have been able to integrate those employed therein.

This task has been easiest, perhaps, with the small STCW companies which are usually owned by individuals who live in the whaling communities, so that both spatial and social distance between owners and employees tends to be close. The owner's personality is thus of great importance to working relations, and an owner who employs a popular gunner with leadership abilities, who keeps his workers on in spite of poor catches, who gives out whale meat liberally, and who does all kinds of small favours for those directly or indirectly in his care, is the most likely person to be able to organize his crew and shore workers efficiently and harmoniously.

A large company, on the other hand, lacks these intimate relations between "owner" and employees. As we have seen, it may well have been formed as a result of a number of mergers and acquisitions, so that there may be sharp divisions between management and employees, on the one hand, and among employees, on the other. To counteract these tendencies, and the fact that larger companies by nature are bureaucratic in their organization, the large whaling companies have followed the example of many other Japanese companies in providing housing and health care for employees and their families, handing out generous lump sum retirement pensions, publishing company newspapers and supporting labour unions and recreational activities. One company (Nissui) has also established its own maritime school.

It is generally conceded that Taiyō Gyogyō is in these respects the least bureaucratic—and hence most "paternalistic"—of the three large whaling companies. For example, although all of them publish monthly newspapers, for both present and past employees, Taiyō was credited for its better information service in that the company regularly sent people all the way to places like Arikawa to keep whalers and their families informed about company events and progress. "As long as my husband worked for Taiyō, we could discuss anything—including finances—related to the company. But once it was amalgated to form Kyōdō Hogei, everything suddenly became very secretive", one Arikawa housewife complained, before continuing: "Taiyō even used to tell us how many whales had been caught". One of the whalers thought that it was Taiyō's labour union which was the reason for this openness. Taiyō had long had its own enterprise union (a typical feature of the Japanese labour movement [Cole 1971; Hayami 1981]), whereas the whalers in Nissui and Kyokuyō were organized as part of the National Seamen's Union (*Zen-Nippon Kai'un Rōdō Kumiai*).[3]

Taiyō Gyogyō has also been more active in sponsoring local branches of its Old Boys' Association, through which retired whalers can continue to meet, "eat from the same cooking pot", and stay in touch with their old companies. These employees' associations both separate people at the lower level, while uniting them at a higher one. In most large whaling companies, the Old Boys Associations are organized according to whether former employees worked on catcher boats, as crew on factory ships and other support vessels, or as processing workers (Taiyō Gyogyō 1984). Each regional branch of the association is then organized hierarchically on a national grid, and this allows members from different specializations—both at the national and local levels—to gather together on certain occasions. In Taiji, however, an all-embracing OB Association was formed in 1982 in reaction to the IWC's decision to impose a world-wide moratorium on commercial whaling, since this was seen as an attack on the whalers' culture and identity. This association includes in its membership people from all companies and from all types of whaling.

Company Rituals

All whaling companies sponsor a series of rituals—both at sea and on land—connected with whaling. Some—such as whale memorial services, or daily offerings at the company's Shintō altar—are performed by the company's management, who will also visit a shipyard for the *funadama* ceremony, which takes place when a new ship is built and its guardian deity is put into place. These sacred symbols are placed in a location of special importance, such as below a catcher boat's harpoon gun.

---

[3] Whalers in Taiyō later joined this union as well, so that when Kyōdō Hogei was established in 1976 all the whalers were members of the same labour union.

Taiyō's paternalistic characteristics may also have something to do with company ownership. Unlike Nissui, Taiyō has been controlled by one family, the Nakabes, except during the first post-war years when the family was temporarily purged by the Occupational Forces, SCAP.

Other rituals are conducted by the whalers themselves.[4] For example, before commencing a journey to the Antarctic, protective amulets or charms (*ofuda*) are fetched from the famous Kompira Shrine in Shikoku, and these are placed on the ships' "god shelves". Individual whalers also visit their local shrines in order to obtain their own amulets before leaving home, and often whalers employed in the same company would go on a pilgrimage to one or more shrines before leaving on a whaling trip. Taiyō Gyogyō, for example, used to bring together whalers and their wives, both from Arikawa as well as from the neighbouring towns of Kami-Gotō and Shinuonome, for a joint ceremony at Arikawa's Sobokimi Shrine, which is dedicated to Izanami, the creator goddess, who—together with her brother Izanagi (enshrined in Shinuonome)—is believed to protect Arikawa Bay and consequently has always been of great significance to whaling in Arikawa. Whalers in other companies (and in other places like Ukushima) would also make pilgrimages with their wives to local village shrines. In Abashiri the Shintō priest still purifies the STCW boats before the whaling season starts, and in Ayukawa each crew takes its boat to the shrine at Kinkazan in order to obtain charms and pray for a successful season. Boat-owners will also take their crews to the local shrine in Ayukawa itself for a brief purification ceremony.

Before a fleet leaves harbour for the Antarctic, a ceremony is held on board the factory ship, with all whalers gathered on deck. This is a time when whalers find themselves listening to speeches by the company's management and singing company songs. Crossing the Equator is an occasion for further rituals. On a factory ship people might be divided into groups according to their place of origin or work assignments, and a big carnival-like party is held, with all sorts of team games being played. The celebrations on the catcher boats are more modest as a large part of the crew at any one time is on duty, but crew members do gather on the bridge for a drink as soon as their duty is over. One of the officers will present sacred *sake* to the boat's "god shelf" and one bottle of such *omiki* is then poured from the boat into the sea. The officers, and those other crew members who feel so

---

[4] The rituals outlined here have in recent years been simplified considerably, while some may have disappeared, but we have retained the present tense to describe them since most of them are still performed (albeit less elaborately perhaps) in connection with the very limited STCW performed in Japan's coastal waters, as well as with the research whaling carried out in the Antarctic.

inclined, then gather to pray for a safe voyage and good catches.

A number of rituals have also been performed during the hunting of whales. Daily ceremonies are conducted in front of the "god shelves", both on the catcher boats and processing units (factory ship and shore station). In times of poor catches or accidents, it has been common to have a "fishing ceremony" (*ryō-matsuri*) on the catcher boat in order to "turn its luck" (*mannaoshi*). The crew offers *omiki* to the "god shelf", and then has a party in which they drink, sing, dance and generally make merry. This party is a kind of ritual designed to improve the crew's concentration and attentiveness, much in the same way as a feudal warrior (albeit in different mood) often used to attend a tea ceremony before he engaged in battle. A similar ritual can still be found in Ayukawa where, in times of poor catches, all those concerned (crew members, relatives, friends, neighbours and others) will gather at the STCW owner's house and dance and sing to the rhythm of drums. Some of the participants may wear carnival costumes, toss the owner in the air and consume large quantities of *omiki* rice-wine which has first been offered to the "god shelf" of the house. The name of the ritual, *taru'ire* ("putting into casks"), probably reflects this desire for a good catch. In other words, it suggests that the casks should all be filled with whale meat.[5]

If a fatal accident happens to occur during the whaling season, the company will see to it that proper ritual action is taken. If possible, a Shintō priest will immediately be called in order to purify the boat and the place of the accident. If somebody is killed in pelagic whaling, the company will provide a temporary funeral on board and a Buddhist altar will be made for the deceased, who is then venerated daily until the ship comes back to port, when the company hands over the remains and the altar to the bereaved family. For several years thereafter the deceased will be remembered in total silence whenever the boat passes the site of the accident. In the past, the company also used to help the family concerned if necessary. One Arikawa widow of a Taiyō gunner was thus assisted in establishing a wholesale business in whale products. It has been through this kind of consideration that employees felt that their company really cared for them.

---

[5] Anybody can take the initiative and suggest that an owner perform *taru'ire*. The owner's wife has to prepare the food and the owner himself provide a lot of alcohol, although the self-invited guests also bring their own supplies of *sake*.

As soon as the season comes to an end, other rituals follow. A company may take all the whalers and their wives to a village shrine, as Taiyō used to do when the whalers returned to Arikawa. These rituals are made in order to show the whalers' gratitude towards the deities for their cooperation in protecting them from accidents and providing them with whales. Whale memorials (*kuyō*) have been sponsored both at local temples where the whalers live, and in temples in Tōkyō where the company management has prayed for the souls of the killed whales.

Through these rituals sponsored by their employers, whalers have been more firmly integrated into their whaling companies. Some of the rituals, such as the ceremonies held before leaving port or marking the crossing of the Equator, have involved most of the whalers in some way or another, and through these rituals whalers from different sections and from different regions of Japan have been brought together. Other rituals have been more localized—either geographically, such as when the whalers go to shrines and temples before and after the season, or sectional, as when the crew on a catcher boat decides to drink to "turn its luck". Finally, there are rituals in which only a few of the managerial staff have participated: for example, when a new boat is launched or when whale memorial services are held in Tōkyō. One common feature of all these rituals, however, is that they are meant to give the whalers a sense of belonging, a sense of security in a dangerous job, and a sense of being cared for.

This point is made more obvious when we consider certain rituals performed by whalers' wives on land. In Arikawa, for example, the latter mark their husbands' crossing of the Equator by gathering at the Sobokimi Shrine where purification rites (*harai*) are performed. On this occasion they bring sacred *sake* (*omiki*) and food with them in order to have a communal meal with the deity. Again, not later than January 10, they gather at the shrine for a four-hour long *miyagomori* ("gathering at the shrine"), with the priest praying for a safe voyage. When the hunt has been completed and the whalers are about to return home, the wives are informed by the company concerned and they again gather at the shrine to pray for a safe journey home. Given that, as we have already seen, wives also accompany their husbands to shrines at the beginning and end of every season, we can see that through rituals the whaling companies served to incorporate not just whalers, but their families as well, into a whaling culture which is both regional and national.

Diffusion of Knowledge

Finally, let us turn to the way in which companies contribute towards an integrated whaling culture through a diffusion of knowledge. The Japanese enthusiasm for learning, of course, is not new and one cannot fail to be impressed by the extent to which technologies were diffused during Tokugawa times (1600-1868)—a period known for its rigid social control and severe restrictions on people's movements. As we have seen in Chapter 6, Fukazawa Gidayū, one of the main operators of whaling groups in Kyūshū, travelled in the 1670s both to Kayoi and Taiji in order to study the net method. Later on, in the 19th century, the Sendai domain tried to develop whaling off Ayukawa by inviting whalers there from Taiji and Yamaguchi, while the central feudal authorities attempted to establish whaling in Hokkaidō by employing people from Katsuyama on the Bōsō Peninsula in present-day Chiba Prefecture (Iwasaki 1987:14-17). People from Fukui were also brought to Hokkaidō to develop the whaling industry, although without much success. In more modern times the founder of Tōyō Gyogyō, Oka Jūrō, travelled to Norway to study modern whaling and for more than 30 years Japanese whaling companies used Norwegian gunners, partly for the purpose of training Japanese.[6] Finally, in the early 1930s, a STCW boat was taken from Taiji to Ayukawa in a successful attempt to modify it for minke whaling.

Today this readiness to learn manifests itself in several ways. For example, in 1989 the owner of one of the Abashiri boats planned to travel to Wadaura to study how to process Baird's beaked whale properly, since he expected this to bring him a higher price for his products in that market. Another company planned to hire flensers from Arikawa to train local villagers if its attempt to open a new flensing station for pilot whales in southwestern Japan proved successful.

These are all examples of whaling operators or leaders taking the initiative in learning new methods of whaling. On the other hand, there have been constraints working against a rapid diffusion of

---

[6] In this context, it is interesting to note that when establishing overseas subsidiary whaling companies in, for example, South America, Japanese whaling companies used experienced workers (primarily gunners and flensers) to teach local employed people how to carry out their tasks. In this way, the companies merely extended a pattern of employment and diffusion of knowledge that they had successfully used within Japan some decades earlier (see Chapter 6).

knowledge—particularly at the lower levels of the organizations concerned. One of the most effective of these has been the importance of having "connections" if someone wished to be employed in a whaling company, since he needed to know somebody who was already hired there. Another related factor has been the extensive use of experienced whalers as migrant workers from the southwest, rather than the education of new whalers in the newly opened whaling bases in the northeast. Seasonal migration can in this way be said to have effectively slowed down the rapid transmission of knowledge of whaling techniques throughout Japan.

These constraints were finally overcome when whalers from southern parts of Japan settled in the emerging whaling towns in the north. By making friends and by establishing real and fictive kinship (i.e. *ebisu*) ties with these settlers, villagers there slowly managed to get the necessary contacts which, as we have seen, helped them find employment in the whaling companies. The shift from seasonal migration to permanent resettlement thus had an extremely important effect on the spread of whaling culture in Japan.

A third, and more permanent, factor which slowed down the learning process was the reluctance of some whalers to teach newcomers their skills. As one of them put it: "To teach your skills to others is to invite unemployment!" Many of the whalers, therefore, guarded their knowledge carefully in order to safeguard their jobs. One informant deplored this situation, since training tends to be done through *minarai*, or "learning by observation", a method much practiced by artists and artisans in Japan who seldom answer questions about their working methods put to them directly by their pupils or apprentices. *Minarai* can thus be seen as a mode of reinforcing the old adage that "knowledge is power", for it effectively stifles the recognition and development of talent (cf. Moeran 1984:159-178), while encouraging a hierarchical organization between senior and junior based primarily on the criterion of length of service.

In order to counteract this somewhat "feudal" mode of operating on board whaling vessels—where individual whalers were reluctant to teach their skills, not for fear of becoming unemployed so much as to be able to monopolize knowledge and hence gain influence—most whaling companies developed their own training methods and facilities. This in turn enabled them to take on and employ local people and so reduce labour costs. Nihon Suisan, for example, ran a training programme for flensers in Onagawa, while Taiyō Hogei had a special "shooting range" where would-be gunners could practice. Whaling companies thus acted as important linking institutions, allowing the dissemination of knowledge to take place in spite of certain inhibiting traditional customs.

But it was the work place—whether on board a catcher boat or in a processing unit—which more than anything else served to transmit knowledge about whaling because it served to bring together people from all over the country. We have seen that whalers were recruited widely during the days of net whaling. Not surprisingly, perhaps, with the development of modern whaling—and especially with the development of pelagic whaling—further steps were taken to bring together whalers from every part of Japan (Appendix 4a), uniting villagers throughout the nation in a way never achieved before.

This broad-based recruitment area has characterized all kinds of whaling. For example, in pelagic whaling, when Kyōdō Hogei's factory ship was sent to the North Pacific in 1986, there were people from 31 prefectures—from Hokkaidō in the north to Kagoshima in the south—plus two foreign inspectors on board. In Chapter 6, we saw that the whalers in the LTCW company Nihon Hogei, particularly those employed on the catcher boats, also came from all over the country. Even the STCW companies have brought together whalers from a variety of villages—witness one of the Abashiri companies, for example, which so far has employed six gunners since 1953. Of these, one came from Kagoshima Prefecture, one from Taiji, one from Ayukawa, and two from Aomori Prefecture, while the present gunner is from Ishinomaki, near Ayukawa. This situation is by no means unique. A number of the other crew members also come from other communities than those in which the boat is registered. (See Appendix 4c.)

By bringing together whalers from all over the country, the whaling companies have thus provided fertile ground for all kinds of transfer of knowledge. But we should not ignore the fact that this recruitment pattern could also cause whalers problems in establishing harmonious working relations among themselves. Indeed, one problem has been verbal understanding resulting from regional dialects, and Nissui, for example, ordered its whalers from Arikawa to teach their dialect to other crew members from the north, since misunderstandings could be fatal in a work team where speed and precision were vital. All the same, cultural differences were probably more important than linguistic ones, and as a result the whaling companies have devoted much energy to fostering a company identity which could unite all their workers.

Conclusion

In this chapter, we have outlined some of the ways in which whaling companies have contributed to the development and maintenance of an integrated whaling culture in Japan. Our focus has been on linkages between companies, which have operated joint ventures as a means of acquiring new skills, reducing financial risk, and adapting to restrictions imposed on them by the Japanese government; on the means by which a company culture has been created and maintained through the performance of various rituals; and on the ways in which whaling companies have contributed towards or, on occasion, acted against a diffusion of knowledge among the country's various whaling communities.

Whaling companies were important because in general they encouraged the breakdown of specialized or secret knowledge within them. Certainly, personal connections were vital in the recruitment of labour and such connections allowed members of a whaler's family to enter into the same company, but they were not necessarily employed in the same set of activities (of hunting or processing) as the relative already employed there. This meant that there could be intergenerational differences in specialized knowledge within the same family, so that, whereas a father would be working on a catcher boat, a son might become a flenser (or vice versa); where an uncle worked as a boiler man, a nephew could be employed as a mechanic on a catcher boat; or brothers might be employed in a wide variety of occupations within the same company. The existence of these different specializations within a single family group, of course, encouraged the free and uninhibited communication of sets of specialized knowledge adhering to each of the two sets of activities (hunting and processing) outlined in previous chapters.

Some activities that we have discussed in this chapter in general create an "in group" atmosphere that effectively separates one company's employees from another's. Whalers will talk about their company—say, Nissui—as though it were the only company capable of carrying out whaling, and they will be more or less ignorant of how other companies (like Taiyō Gyogyō or Kyokuyō Hogei) are organized. Although whaling companies encouraged the breakdown of secret knowledge within their own organizational structure, they simultaneously tended to *create* a cultural knowledge, too, that was partly specific to each company. Thus, in some respects, they became exclusive entities, while simultaneously "rationalizing" knowledge and thus advancing their own effectiveness in a competitive industry. At the same time, however, we have mentioned certain linkages between the companies that should not be ignored.

Given that whaling companies clearly have contributed to the formation and maintenance of a truly integrated national whaling culture, we now need to ask what has been the effect of their having been obliged first to cut down on, and then cease entirely, their whaling activities in response to the international ban on commercial whaling. This is a question which has been implicit in much of our discussion in previous chapters on work organization, recruitment patterns, and local whaling culture. It is now time to address directly the question of the overall effect of the moratorium on all types of commercial whaling in Japan. This is the subject of our concluding chapter.

CHAPTER 9

The Impacts of the Moratorium

We have elsewhere defined "culture" as an integrated and coherent system of specific tools, techniques, skills, and the attendant bodies of knowledge and forms of social organization that are necessary to locate, identify, harvest, process, distribute, and consume particular resources found in specific ecological niches (Takahashi et al. 1989:105). In Chapter 5, we analysed the three types of whaling found in Japan—i.e. pelagic whaling, LTCW, and STCW—and found that similarities in tools, techniques and skills required in these forms of modern whaling were striking. There are nevertheless, within each type of whaling, marked discontinuities between the two sets of activities related to hunting and processing respectively, together with certain minor disparities between the various forms of whaling. These discontinuities and disparities have forced us to investigate possible linkages which have served as bridges between the two sets of activities and between the three forms of whaling discussed in this book. In Chapters 6, 7 and 8, we looked at the diffusion of knowledge and the social organization facilitating such transfers of knowledge. More precisely, in Chapter 6, we saw how, through an analysis of recruitment and career patterns, knowledge travelled from one form of whaling to another and, in our study of a kinship group, from one set of activities to another. Then in Chapter 7, we focused on the role of the local whaling community: on how whalers, regardless of their respective tasks in the whaling operations, were brought together through gift-exchanges, rituals, belief systems, and unique food preferences, thus giving rise to what we termed *local* whaling cultures. At the same time, however, we saw in Chapter 8 that these local whaling communities were then linked together by the whaling companies. It is these many linkages in both communities and companies that have enabled us to talk about a *national* integrated whaling culture, bringing together the two sets of activities and three forms of modern whaling outlined in this book.

This whaling culture is, of course, not static, but continuously in the making. Japanese whaling culture has faced many challenges in the past, and it has survived, albeit in changing forms, only by making certain necessary adjustments. In recent years Japanese whaling has again faced a challenge of great magnitude—a challenge which strikes at the very heart of whaling itself: the moratorium on commercial whaling forced upon Japan by western powers. Is this the ultimate challenge which will put an end to a centuries-long tradition?

In this concluding chapter we will take a closer look at the economic, social and cultural impacts of the steadily dwindling quotas and of the subsequent total moratorium on the whalers and their communities. In other words, we wish to address the problem of what Japanese whalers themselves regard as an endangered culture.[1]

Whalers and Their Families

When the moratorium on commercial whaling was first enforced a number of whalers were laid off work. Of the 507 persons employed by Nihon Kyōdō Hogei in Antarctic whaling, only 369 were rehired by the new company Nihon Kyōdō Senpaku which now conducts

---

[1] We will not discuss possible ecological consequences of the moratorium here as we feel that this lies outside our competence. There are reasons to believe, however, that a total moratorium on hunting of sea mammals will seriously harm the ecosystem unless a moratorium is also imposed on fishing. Since toothed whales feed on fish and molluscs and baleen whales seem to do so to an extent not hitherto realized (see Mønnesland et al. [1990] for minke whales' consumption of fish in Norwegian waters), the moratorium on whaling coupled with commercial fisheries will most likely bring about ecological imbalance with the predators (man, whales, and seals) ending up starving.

An important point to be made here concerns environmentalists' "Save the whale" campaign. Given the way in which fishing resources as a whole are affected by the proliferation of some species of whales, we need to ask precisely for what the whales are being saved. If certain types of whale were to become too numerous, their food supplies would dwindle and they could literally starve to death (see Terhune 1985 and Aron 1988 for an appraisal of the situation). Rather than call for a blanket moratorium on all commercial whaling, therefore, we need to ensure that a successful management system is devised to ensure an "ecological niche" (Barth 1956) for each species of whale, whose habitat can then be secured, which incidentally is also the view of Japanese and other whalers who argue that it is necessary to harvest the ecosystem at all levels in order to preserve the ecological balance. (See also Manning 1989:227.)

research whaling on behalf of the Institute of Cetacean Research. The three LTCW companies Nihon Hogei, Nittō Hogei and Sanyō Hogei laid off 282 full-time and 42 part-time employees. All but one crew were laid off by the eight STCW companies, but about half have since been rehired in order to catch beaked and pilot whales. The laid-off whalers have had serious trouble in finding new employment, and in order to understand why so few have secured permanent jobs, we have to look both at the availability of work locally and at the skills whalers bring with them to the labour market.

Given that the largest of the old whaling companies—Nissui and Taiyō—have also been heavily involved in fishing, critics of Japan's whaling industry might justifiably argue that former whalers can be redeployed in Japan's fishing industry. After all, Japan is the world's biggest fishing nation by volume of catch, and—until recently at least—fish and shellfish have provided half of the animal protein requirement in the Japanese diet. In 1989, its fishing industry employed about 440,000 fishermen, but the worldwide trend towards the establishment of 200 mile exclusive economic zones has created a critical situation for the industry, which used to rely on fishing within the 200 mile zone of other nations for nearly half its annual catch. Fishing concerns have tried to compensate for this loss by expanding their trading activities (Stokke 1991).

As a result of such measures, Japanese fishermen have lost access to a number of fishing grounds, while being obliged to pay fees for the use of others. This in itself is to be expected in a free market economy, but the fishing industry has also in the 1980s been plagued by continued pressure and threat of sanctions by the United States because of Japan's refusal to give up whaling. Not surprisingly, perhaps, many whalers feel that they have been sacrificed to their country's fishing interests, but we should note that the Japanese government is also in the process of buying back fishing licences in order to get companies to scrap their pelagic vessels. Consequently, it is virtually impossible in such economic conditions for whalers to find employment in the pelagic fishing industry.

The situation is no better in the coastal waters of Japan where the management of resources has been carried out over many centuries by local bodies (villages and, after 1901, fishing cooperative associations, or FCAs). As mentioned in Chapter 7, one of the main features of coastal fishing in Japan has been the village-based exclusive fishing territories administered by FCAs. Only members of the FCA are in general entitled to fish within the territories administered by that association. Furthermore, a person usually has to be a resident of the village—as well as an active fisherman—in order to be a member of the FCA. This means two things: firstly, fishing is not an open frontier to the landless in Japan, as it has been in some western countries (cf. Löfgren 1979:102); secondly, whalers find it difficult to become members of their local fishing cooperative unless they live in the fishing community concerned and fish actively.

The ability of whalers to take up coastal fishing is further restricted by two important problems: those of space, and of licences allocated by the FCAs. Licences are in very short supply and have in most cases already been allocated. This means that newcomers can only use those forms of fishing technology for which licences are not required—in other words, the least productive methods such as hand-line fishing, jigging, and seaweed gathering. It is precisely in these fisheries—which are little more than subsistence activities—that we can find some former whalers.[2]

Fishing activities also imply usage of sea space, which is a particularly critical factor when technologies, like those used in aquaculture and stationary nets, monopolize space for an extended period of time. Although large set nets (*teichiami*) were first invented in the Tokugawa period and were extensively used, for example, in Ayukawa during the late nineteenth century where a new village hierarchy emerged as a consequence (Chapter 2), new and larger types of nets have, during the last few decades, spread rapidly throughout Japan. Indeed, they have become so popular that they occupy large areas of sea space for months at a time, thus causing considerable inconvenience to other kinds of fisheries. More important, many fishing territories have little space left to locate new set nets. In Arikawa, for example, ten of these nets were in 1988 operated by a cooperative of fishermen, Arikawa Gyogyō Kyōdō Keieidan (AGGK), which in the past has agreed to recruit former whalers provided that the whalers were under 50 years of age (Kalland 1989:127). However, AGGK has recently been unable to introduce new nets because of lack of space, and it has therefore been unable to take on even young whalers (Map 3).

---

[2] Of about 100 handliners in Arikawa FCA in 1989, for example, 37 were former whalers. Only a few of these handliners earned more than ¥3-400,000 per year. The situation is similar in other villages.

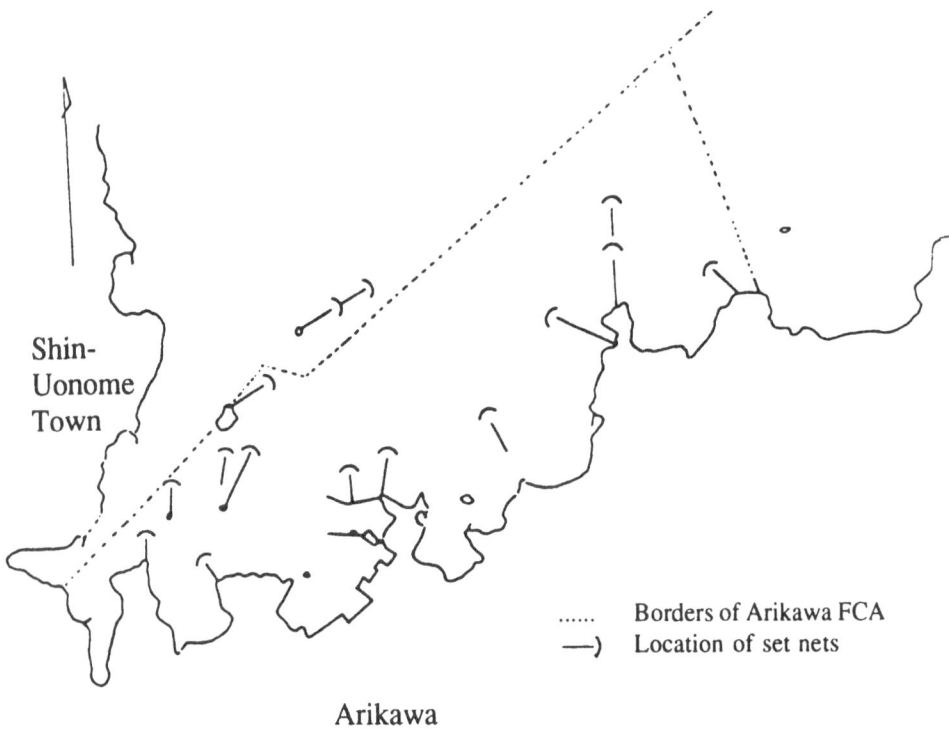

Shin-
Uonome
Town

...... Borders of Arikawa FCA
—) Location of set nets

Arikawa

Map 3:	Location of stationary winter nets within the fishing
territory administered by Arikawa FCA.

Aquaculture has been heralded as the industry of the future in
many maritime communities in Japan. But, in addition to the general
shortage of space in local fishing territories, there are serious
constraints for large-scale aquaculture in most whaling communities,
which are usually located in remote and rocky peninsulas and islands,
where rough waters often make aquaculture both expensive and
hazardous. Pearl oysters are grown inside the calmer inlets in Arikawa
and Taiji towns, and silver salmon (*ginzake*) are bred in Ayukawa,
where most of the STCW companies have taken to salmon breeding
to make ends meet. But, as with many new ideas for making money
in Japan, too many people have leaped upon the bandwagon of
ginzake breeding. This has led to problems of pollution, on the one

hand, and to overproduction, on the other.[3] The latter, when coupled with an increase in imports of salmon from Norway and Canada, has led to deteriorating prices which have now reached a level that makes it impossible for the Ayukawa companies to make money, and by January 1992 one of the operators had gone bankrupt.

Only a few of the whalers have been employed in fish breeding in Ayukawa. The new work is unfamiliar and they have had great difficulty in getting used to it (cf. portrait of Abe Yoshio in Chapter 3). After all, the whalers are not fishermen and are not used to working on small fishing boats. This is, of course, particularly true for flensers, operators of oil boilers, and those who worked on refrigerator ships.

Given that many of the whalers originally came from farming households, it could be argued that they might be more familiar with farm work than with fishing. But the environment in which most Japanese whaling communities are situated also severely restricts opportunities for agriculture, for there are hardly any extensive areas of fields near any of these towns. In Arikawa, for example, only 4.3 per cent of the land area is farm land (Kalland 1989:114). In Wada, the percentage approaches the national average at 20.9 per cent (Wada-chō 1987:2), but in Oshika it is only 4.6 per cent (Oshika-chō 1988:97). Moreover, the small patches of farmland that do exist are often of poor quality and are thus best suited for dry-land crops.[4] It has thus not recently been possible to live off the land alone, and farming has had to be combined with such by-employments as whaling, fishing, forestry or seasonal migrant labour (*dekasegi*). In Arikawa, for example, it was income derived from whaling which enabled many households to carry on from one generation to the next and thus continue to farm their land. However, with the collapse of whaling and with few alternative sources of by-employment, families have been forced to move from Arikawa and leave their land untilled. At least in this town, therefore, agriculture has declined partly as a

---

[3] Even so, there is a Government project which has selected Oshika as one of three places in the country where bays will be closed off by long breakwaters in order to improve conditions for aquaculture. The project has caused considerable local controversy of both an environmental and political nature.

[4] In Arikawa rice comprises only 4 per cent of the value of agricultural production, while in Oshika and Wada the area suitable for wet rice cultivation comprises only 1.3 per cent and 13.7 per cent respectively.

result of the declining whaling industry.[5]

There is also little help offered by the authorities to develop agriculture in these remote areas. Indeed, the Japanese government—in response to international pressure—has been trying to reduce (rather than expand) domestic agricultural production in order to open the country's market to imports. As a result, many farmers, including some Arikawa whalers who have invested their earnings and time in developing orange orchards, have been obliged to cut down their trees in order to help open the Japanese market to imports of fruit from the United States.

With whaling gone, the farmers may rely more on forestry, manufacturing industries or services to supplement their meager incomes from the land, but there are serious constraints here, too. The hills are steep and exposed to erosion, making forestry difficult and risky, and only a handful of loggers have found employment in the forests.

As we have already pointed out, Japanese whaling communities tend to be located in remote parts of the country and Ayukawa, Wadaura, Taiji, Ukushima, and Arikawa are all located on rugged peninsulas or islands. They are far from central markets; transportation is slow and expensive; and it is very difficult to persuade industrialists to invest in these areas. The only exceptions to this are polluting, and potentially polluting, industries like nuclear power plants and petrochemical industries which, as we have seen, can be found in the vicinity of Ayukawa and Arikawa respectively. It is also difficult to make a career in the tertiary industries. Although the number of people engaged in this sector has been on the increase until recently, there are indications that this trend has come to an end. In fact, the number of shops is on a decline in several of the local towns (see below).

There are thus not many alternatives for work available to whalers who are ill prepared to compete for the few work openings that do exist. This is because firstly, they—particularly gunners, bosuns,

---

[5] The decline in agricultural output has been dramatic in Arikawa Town. Whereas 403 tons of rice were harvested from 112 hectars of rice fields in 1960, only 59 tons were harvested from 22 hectars in 1980. The harvest of wheat declined from 579 tons to 9 tons, that of sweet potatoes from 4,796 tons to 541 tons and the amount of potatoes from 496 tons to 330 tons. On the other hand, the area planted to *amanatsu* oranges has increased from nil to 72 hectars, which produced 800 tons in 1980 (Arikawa-chō 1982:233).

flensers and operators of boilers—possess skills not much sought after outside the whaling industry. Captains, engineers and cooks are in a better position to find employment in the merchant navy, but even these have experienced difficulty in finding new jobs (Government of Japan 1989:10-11). In general, therefore, it is probably fair to say that the laid-off whalers have found their skills to be both irrelevant and useless.

Secondly, most of the laid-off whalers are in their 40s or more, and are as a result not very attractive to would-be employers. Mid-career recruitment has until recently been difficult in Japan (and even now is only practiced in certain city oriented businesses), and it is very hard for such people to find a good, permanent job which carries both prestige and such fringe benefits as a pension and health insurance. At best, they are able to get temporary employment with greatly reduced salaries and little social security.

Finally, we know that in many whaling communities it was regarded as a personal failure for someone not to have been employed in whaling, so that whalers have in general been used to working in a prestigious occupation. Many have thus found it degrading to seek not just temporary work (itself of low status), but work outside the whaling industry, since such work tends to be less prestigious. Indeed, those accepting such work are often looked down upon as being near the bottom of the ladder of social status, so that, in order to avoid facing social humiliation, some of the older whalers have not been motivated to take up any employment at all and have thus remained unemployed. But in a society which puts so much emphasis on work and where people get their social identity through their work, it is also very embarrassing to be without a job and to receive unemployment compensation. Unemployed whalers thus not only suffer a loss in prestige, but are also partly marginalized.

Whatever they do, most of the whalers find themselves with a greatly reduced income, although most of the full-time whalers laid off from large type whaling were entitled to unemployment compensation for up to one year. Whalers used to be among the more affluent inhabitants of their villages, and many managed to get their children into good schools.[6] Their affluence also enabled them to extend their

---

[6] As a matter of fact, whaling seems to have been a path to upward social mobility as many whalers were able to convert relative wealth into education for some of their children, who then have made a career for themselves in the cities.

social network very widely by entering into extensive exchanges of purchased or money gifts, in addition to giving away whale meat. With reduced income, of course, the whalers are no longer able to live up to such expectations, nor are they able to carry out their social responsibilities as household bread-winners and participants in social networks. Many of their wives, too, have been forced to take up employment in order to support their husbands, but this upsets accepted gender roles in Japan and so further humiliates the husbands.

All these difficulties have led to several of the whalers suffering from psychological problems, and stress-related disorders seem to be on the increase (Government of Japan 1989:12). In Ayukawa one laid-off whaler has died of excessive drinking and in Taiji another was hospitalized for the same reason. Down in Kyūshū, a third whaler is said to have gone mad when he chopped off one of his fingers. These dramatic cases are only the tip of the iceberg. Many whalers and their wives suffer from stress and feel anxious about the future, and cases of neurosis have also been reported (Takahashi 1988; Government of Japan 1989:12-13).

Demographic Trends and Village Economy

With these difficulties in finding re-employment it is not surprising that the populations in most of the whaling towns (i.e. in Wada, Oshika, Ukushima, and Arikawa) are declining rapidly. Only in Abashiri and Taiji have they remained rather stable (Table 4).[7]

The population trend has also brought about other demographic changes. In the first place, for example, the decline in population has not been matched by a similar decline in the number of households. There has been a modest decline from 1,898 to 1,840 in Wada between 1960 and 1985 and a somewhat steeper decline in Ukushima from 2,435 to 1,972. On the other hand, in Ayukawa the number of households remained stable at about 800 during this same period—but has since declined to 745 (1990)—while Arikawa saw an increase of

---

[7] The population figures of the townships conceal internal movements within municipalities, however. In Arikawa Town, for example, there has been a marked centralization in that people have moved from outlying hamlets to the centre (Kalland 1989:115-116). In Oshika Town, on the other hand, all the hamlets have been hit by depopulation more equally.

Table 4: Population Trends in Abashiri, Oshika, Wada, Taiji, Ukushima and Arikawa

	Abashiri	Oshika	Wada	Taiji	Ukushima	Arikawa
1940	32,732	9,902		3,570	9,154	8,903
1947	34,850	11,798		4,346	11,401	11,058
1950	39,218	13,226	10,339	4,656	11,466	12,208
1955	42,961	13,753	9,977	4,591	11,684	13,202
1960	44,052	13,405	9,108	4,556	11,175	13,280
1965	44,195	11,974	8,396	4,605	9,503	12,018
1970	43,904	10,581	7,761	4,566	8,048	10,806
1975	43,825	9,535	7,291	4,433	6,689	10,058
1980	44,777	8,450	6,879	4,539	5,840	9,882
1985	44,285	7,814	6,567	4,314	5,222	9,393
1990	42,733	7,323	6,243	4,098	4,808	8,551

20 per cent from 2,514 to 3,027—but has since been stable. This implies that there has been a steep drop in the average household size in these townships. In all the townships, with the exception of Wada, the average household size now falls well below the national average (which stood at 3.01 in 1990).[8]

Secondly, if we examine the age distribution of the populations concerned we will notice that there are relatively few persons between the age of 15 and 25 years in the whaling towns (Fig. 3). One reason for this is that many young people have to leave home in order to go on to high school (and university) education, since the remoteness of many whaling towns prevents them from having their own high schools. Children from Ayukawa, for example, have to commute almost as far as Ishinomaki in order to attend the nearest high school. Not surprisingly, perhaps, many of them prefer to find lodgings there rather than make the one and a half hour bus trip twice a day.

Lack of work opportunity is also a serious problem, so that more and more of these young people choose not to return to their natal villages after completing their education. Old people, on the other hand, tend to remain in the communities, and with the Japanese now having the highest life expectancy in the world it is likely that this

---

[8] The figures for the six townships introduced in Chapter 2 were in 1990 as follows: Abashiri 2.68; Ayukawa 2.75 (as against 3.15 for the township as a whole); Wada 3.37; Taiji 2.65; Ukushima 2.48; and Arikawa 2.84.

uneven age distribution will be further distorted in the near future. Of course—and this is a third demographic effect—many of these old people live alone without support from working relatives, and therefore become more dependent on public support. At the same time, given the declining population as a whole, there are few people to pay for this support through taxation. In Oshika, for example, the number of people aged 65 or over has increased by 75 per cent between 1960 and 1985, but the actual working (and hence the tax-paying) population (i.e. those between 20 and 65 years old) has decreased by 30 per cent.

Figure 2: Age Distribution in Oshika Town, 1955 and 1985

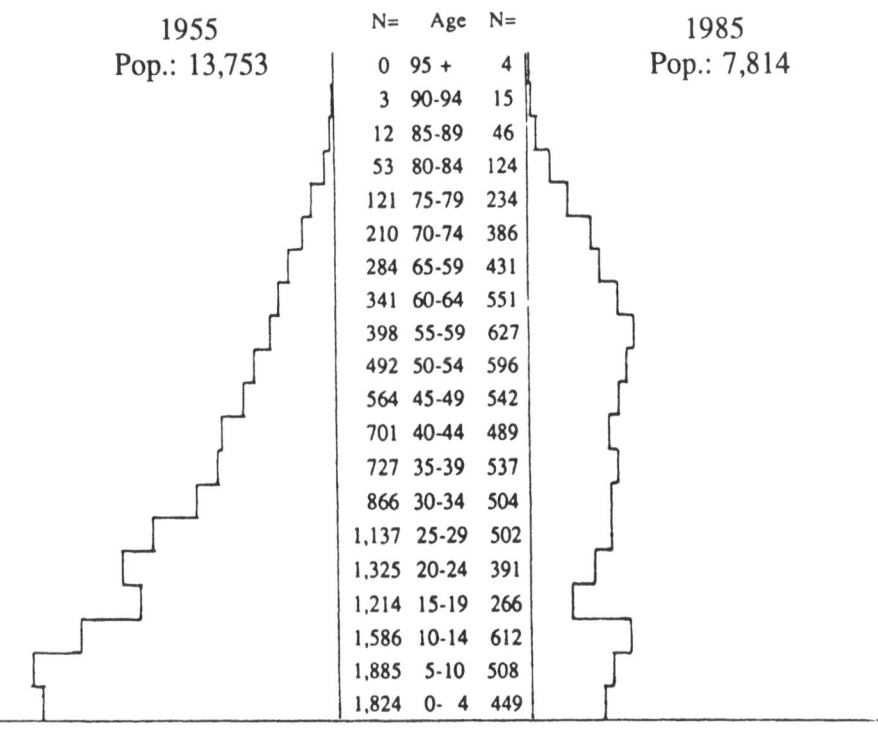

1955 Pop.: 13,753	N=	Age	N=	1985 Pop.: 7,814
	0	95 +	4	
	3	90-94	15	
	12	85-89	46	
	53	80-84	124	
	121	75-79	234	
	210	70-74	386	
	284	65-59	431	
	341	60-64	551	
	398	55-59	627	
	492	50-54	596	
	564	45-49	542	
	701	40-44	489	
	727	35-39	537	
	866	30-34	504	
	1,137	25-29	502	
	1,325	20-24	391	
	1,214	15-19	266	
	1,586	10-14	612	
	1,885	5-10	508	
	1,824	0- 4	449	

It would be wrong to claim that this situation is unique to whaling towns, for it is a problem found in many rural areas of Japan. The point is that whaling was one important means by which an overall social and economic problem was being combatted, so that the internationally imposed moratorium has had further unanticipated socio-cultural consequences. When whaling was commercially viable, it supported a number of other activities, so that the closing down of the industry has had extensive effects far beyond the whaling households themselves.

With the end of LTCW, several businesses in Taiji, Wada and Ayukawa suffered financial losses and many family enterprises went bankrupt. This depression also affected the tourist industry which, incidentally, has often been regarded as the industry of the future for these towns. Some of the tourist facilities—such as inns, bars and restaurants in Wadaura and Taiji—have managed to survive by catering partly to migrant whalers during the tourist off-seasons, and they would be unable to stay in business without these whalers (Government of Japan 1989:27-29). The decline in population furthermore undermines the economic base of local shops and services so that inhabitants will almost certainly become more dependent on outside services in the future.

Further economic repercussions affect the fishing associations of Ayukawa, Wadaura, and Taiji which have lost important sources of revenue. In Ayukawa, the FCA's revenues from whale related activities in the period 1984-1987 amounted to an average of 61.9 per cent of the total revenues from marketing sea products (with a peak of 78.8 per cent in 1987) and 53.5 per cent of the revenues from sale of ice (74 per cent in 1984). 43.4 per cent of the revenues from freezing operations in 1986 and 1987 derived from whale meat (Bestor 1989b). The loss of these revenues has placed the Ayukawa FCA in a difficult economic position, and plans are under way to reduce the number of employees by three and reduce services to its members.[9] The FCA may also become more dependent on rent paid for sea space by operators of fish breeding plants and set nets, as well as on

---

[9] Beside being a crucial institution in the orderly management of marine resources, the FCAs extend a number of services to their members, such as cheap credit and saving accounts; discounts on fishing gear, fuel, and ice; marketing of their products; educational programs for the fishermen, their wives and children; recreational activities; and so on.

compensation money paid by a recently established nuclear power plant and by others who harm the environment on which the fishermen depend. The situation is less critical in Wadaura and Taiji, but there, too, the FCAs had considerable revenues from whale related activities through sales of ice, use of storage facilities, rent of land and shorelines, and so on. In all these communities, loss of revenue to the FCA has meant an automatic decrease in contributions to the communities' general welfare.

The decline of commercial whaling has influenced the markets in other ways as well. Marketing networks have changed, with more of the meat from small whales landed along the coast being sold through national markets as wholesalers do their best to satisfy a non-local demand and attempt to make up for the losses of whale meat from the Antarctic. This, of course, affects local food consumption and makes it seem to outsiders that STCW is not a locally based but a national commercial industry. Moreover, with the moratorium on coastal minke whaling, a completely new situation emerged for STCW in 1988 when most of the Baird's beaked whales were landed in Abashiri, and not in Wadaura as hitherto. This seriously disrupted the marketing structure for this species of whale—a point to which we will later return. As a result of these changes in marketing of whale products, small local dealers have been forced out of business (Bestor 1989b:19-23; Government of Japan 1989:25-26).

All this has, of course, serious consequences for the economy of the towns. With fewer people in the work force and with less industry and trade, the tax base has—as we have already suggested—been seriously eroded. It is not just whaling households, therefore, that suffer financial stress. We saw in Chapter 2 how the once healthy economy of Taiji—where demographic changes have been less dramatic than in most of the other whaling towns—has been turned into a huge deficit. All whaling communities have in recent years been forced to rely more on state subsidies (which undermine their autonomy) and on compensation money from potentially polluting industries. At the same time they have been forced to curb what would normally be regarded as essential services. Educational facilities, in particular, have suffered as local governments find themselves unable to improve school facilities for the steadily declining number of children. With the general importance of education in modern Japan (Rohlen 1983), it is not surprising to find that parents find themselves forced to send their children to nearby towns and cities to attend senior high schools. Again the situation seems to be most difficult in Ayukawa where the nearest high school is located about 40 km away. It would seem from interviews with informants, indeed, that more families move from Ayukawa in order to get their children into schools than for employment. The whaling communities are thus trapped in a vicious circle, a fact which also has implications for the cultural well-being of those who live there.

Cultural Implications

Throughout this book we have focused on "whaling culture" as consisting of systems of shared knowledge. We have analysed at length what this knowledge consists of and how it is transmitted through time and space. It therefore follows that without whaling there can hardly be whaling culture—a truism which at first glance would not seem to need investigating except that, as social scientists, we need to examine *how* the whaling culture is collapsing and *what* will come in its place. This, too, is an area of investigation that should be of vital interest to bureaucrats and policy makers throughout the world.

First of all, knowledge directly related to hunting (in particularly, gunners) and processing of whales will be lost. This process has already started as it has become difficult to find flensers able to cut large baleen whales skilfully. Since most of the laid-off whalers are middle aged or older, it is only a question of a decade or two before the knowledge accumulated by these whalers will be lost, unless younger whalers can be recruited and trained in flensing skills.

But it has for some years been difficult to recruit young people into whaling because of the reduced scale of operations, and with many young people between 15 and 25 years of age being away from the villages, their socialization into whaling culture has been seriously affected. Low quotas and short seasons have also made the training process more difficult. Moreover, the introduction of the hot harpoon in STCW—which was done in order to silence those critics who claimed that whales were killed inhumanely—has made it more difficult to train new gunners. When the cold harpoon was used it was often necessary to fire a second shot in order to kill the whale, and an apprentice gunner was allowed to do this as part of his training practice. With the new harpoon, however, no second shot is necessary, and it has become more difficult to train gunners since the harpoons are expensive and the whalers can ill afford to miss the target.

Being a whaler involves more than just mastering technical skills, however. There is a whole body of knowledge pertaining to the ecosystem which needs to be studied. The whalers—and the gunners in particular—have an extensive knowledge about whale behaviour, about their feeding, migrating, and mating habits. Their perception of the ecosystem is based on a paradigm which differs from that of the marine-biologists, and the loss of this understanding is—in our opinion—unfortunate. Closely related to this question is the whalers' perception of nature in general. Japanese whalers share with many hunters everywhere a world view in which man and nature are seen to interact. This reciprocal relation between the two is very different from the Christian perception that nature exists for man to exploit to his benefit. As far as the Japanese are concerned, however, the whale gives itself to the whaler who in return is obliged to protect the ecosystem and perform memorial services for the whale's soul.[10]

With the moratorium in force, there is no longer any need to go to temples in order to pray for the souls of killed whales or to shrines to give thanks for good catches and for whalers' protection at sea. Whalers' wives, too, do not need to go on pilgrimages or pray for good catches and safe voyages in front of their house altars, and the morning rituals are considerably shortened. Old village festivals are also affected. Some festivals have disappeared, while others have changed in character and survive only on a greatly reduced scale. This is the case, for example, with the once famous *Meizaiten* festival in Arikawa (cf. Chapter 2). No longer able to hunt whales, the whalers find that their perception of and respect for the environment is thus slowly being changed and eroded. One community's Buddhist priest was of the strong opinion that people have lost some of the respect and sense of interdependence which they had previously felt toward nature, although this is probably a national trend to some extent.

Religious beliefs and rituals are everywhere important elements in shaping people's identities, and in Japanese whaling communities

---

[10] We should perhaps add that the perception of a relationship of reciprocity between man and nature among the Japanese depends on a situation where they both belong to the same ecosystem. The Japanese have little concern for nature in some distant part of the world like, say, Africa. In this their relationship towards the environment resembles that which they adopt towards people.

beliefs and rituals have been strongly influenced by whaling activities. Equally important, perhaps, are food preferences, which again are closely connected with the activities of whaling. Thus the loss of whale meat not only means that a rich culinary tradition is disappearing, but that people's identities are being affected as well.[11] As we have argued in Chapter 7, it is precisely religious beliefs, rituals, and cuisine which have given the whaling communities their distinct character and identity. Without whaling these communities are already being forced to find other symbols through which their identities can be expressed, but this is a painful process which makes heavy demands on community solidarity in order to achieve success. Yet, as we have seen, this solidarity is already strained for other reasons.

Another adverse influence on the continuation of whaling culture stems from the demographic changes outlined above. We have seen that the mean household size has decreased dramatically since 1960—something which has serious implications for village institutions. In Chapter 7, we argued that the Japanese village (*buraku*) consists of a number of households, rather than of individuals as such. However, individuals represent their households in a number of village-wide associations—such as farming and fishing cooperatives, children and youth clubs, the PTA, fire brigade, housewives'

---

[11] The importance of whale meat for people's wellbeing is recognized when the meat of minke whales caught in the Antarctic is distributed. One aims at a proportional distribution between the prefectures based on the average of the past four years regional consumption of whale meat (Government of Japan 1989:64-67). Separate amounts of meat are also given to treat food-allergic infants and to Oshika Town.

In 1989 Oshika Town was allowed to buy 10 tons of red meat and 3 tons of scraps. 8 tons of the red meat were distributed equally to all the households in the town. Representatives from the households were invited to buy tickets entitling them to collect 3.75 kg of meat at a designated place and time. 98 per cent of the households made use of this offer. The lodges and restaurants shared one ton of the red meat and the remaining ton was kept by the town office for special occasions when it is important to present an image of the town to outsiders. The town official in charge also contemplated using some of the meat at a match-making party the township had planned on behalf of all its bachelors. The 3 tons of scrap meat were shared between the two private hotels (500 kg each); the public lodging house (*kokumin-shukusha*), hospital and school kitchen (one third of a ton each), and the Kinkazan shrine and the ferry terminal (500 kg each).

In 1992 the meat is distributed in a slightly different way. Each household can buy 5 kg red meat twice at a price of ¥3,000 per kg (subsidized by the town), the school will receive 400 kg, the public lodging house 670 kg, and the FCA 2,270 kg.

association, old people's club, and so on. It is precisely through participation in such associations that households are integrated into the village so that the latter becomes a "community" as such and not just a haphazard collection of independent households. It is obvious that a small household with only a few members (often limited to a particular age group) can participate in only a few of these associations and that such a household can thus become marginalized fairly easily. With many such small households in a village, village solidarity might then be at stake.[12] For example, relatively few households can now participate in the activities of the youth associations or fire brigades (which, therefore, suffer a decline in their memberships). Since these associations play a central role in village festivals, the festivals themselves are also affected.

Also many of the whalers' wives now participate less in village affairs than they did in the past. As we have seen, when their husbands went away on whaling trips, women used to come together regularly for pilgrimages to shrines and temples. The wives in Arikawa, for example, went on two monthly pilgrimages during their husbands' absence. Moreover, they met regularly at the local shrine for ceremonies, and many wives accompanied their husbands to these shrines before and after the whaling seasons. In some of the whaling villages, too, women were put in charge of the fire-brigade. With their husbands at home, however, they have fewer opportunities to associate, since they have to attend to their laid-off husbands or to work to make ends meet. Many women lament this situation, which they feel is breaking down the strong solidarity which once existed among the whalers' wives.[13]

---

[12] In Oshika Town there were 426 one-person households in 1985, 926 households of nuclear families (of which 343 consisted of a married couple only), 849 extended households and four others. This means that at least one third of the households had only one or two members. In Arikawa 22 per cent of the households were one-person households in 1990. A high proportion of the households were thus not able to participate fully in village affairs.

[13] No doubt, westerners believing in the values of romance will regard the fact that spouses are now able to stay together the whole year as an improvement, but many Japanese view the situation in a somewhat different light. For them, marriage is not so much based on the concept of love as on a concept of complementary roles between husband and wife. It is the latter's task to run the house and bring up the children, while her husband provides her with the financial means to do so. The ideal husband is, according to a Japanese proverb "a man who is away from home and

With fewer whales caught, the competition for scarce meat has increased, and prices have soared accordingly,[14] leading as a result to more of the meat being channeled through the national rather than just local markets. The amount of meat available locally has therefore decreased markedly, with serious consequences for the kind of social networks that we outlined in Chapter 7. Local dealers deplore that they have been forced to sever ties with other dealers with whom they had maintained close ties for decades (Bestor 1989b:21). Whale meat is no longer available for gift-exchange—the "glue" of community solidarity—and, with their income reduced, former whalers have no means to substitute whale meat with other products, not can they keep up the monetary exchanges which also used to flourish in these communities. With gift-exchange being greatly reduced in scale, this glue by which social relations adhere is weakened and people's social networks seriously contracted.

So, solidarity is breaking down and former friends and allies have become competitors and, in some cases even enemies. In Abashiri, for example, communication between an STCW boat owner and his laid-off crew has broken down completely and has been replaced by bitter feelings of hostility. In Wadaura one of the processors of dried whale meat (*tare*) cornered much of the market in 1988 and thus forced others out of business. This came about because of changes in the distribution system. As long as beaked whales were landed in Wadaura, fixed percentages of the meat was distributed between local processors, but in 1988 when the whales were landed and flensed in Abashiri and Ayukawa before being transported to Wada, the largest processor did not feel constrained by local Wada customs and purchased the whole lot. With the subsequent closure of the flensing stations in Abashiri, more Baird's beaked whales are now landed in Ayukawa, from where much is sent to Kyūshū, thus creating a shortage in the traditional market at Wada.

If the moratorium is extended *ad infinitum* the outcome is obvious. The situation in these villages will detoriate further, although eventually they will manage to adapt and return to a kind of Japanese

---

healthy". When whalers used to be wealthy as well, they thus made extremely attractive husbands.

[14] The average wholesale price of whale meat at the market in Ayukawa, for example, doubled betwen 1984 and 1987 (Bestor 1989b, table 3).

tranquillity. They will be very different communities, however, with new symbols and identities, and with greatly reduced populations. Many will be obliged to leave home for the already congested life in Japan's cities, and in the process their valuable knowledge of ecosystems and of the utilization of renewable resources will be lost.[15] In this respect, the outcome of the moratorium will work directly against core elements of the World Conservation Strategy (WCS), also endorsed by the Brundtland Commission's report *Our Common Future*, which states that: (1) human life depends upon sustainable and equitable utilization of environmental resources; (2) governments need to promote sustainable development so that they can assure the sustainable livelihoods of all peoples; and (3) governments should recognize and actively involve people at the community level both in developing and implementing sustainable development strategies (Freeman 1990). That the moratorium works to the contrary is another irony arising from the activities of animal rights groups and brings us back to where we started in Chapter 1: the politics of whaling.

Whaling and Nationalism

Why whales? We addressed this question briefly in the first chapter of this book and will not repeat ourselves here. What we are after now is in fact not an "objective" scientific analysis of totems and the like, but the "native theory". After all, we have already subscribed, it would seem, to anthropologists' tendency to romanticize those cultures they study, by referring to whaling as an "endangered" culture. We should make it clear therefore, that the feeling that Japanese whaling has come to the end of an era, and that its culture is endangered, is not just ours, but that of the whalers as well. Time and time again they, too, have asked themselves "why whales?", but have yet to find a rational answer.

---

[15] A further argument can be made along the lines that the demand for labour in such export-oriented industries as automobile manufacturing allows those hitherto employed in the whaling industry to take up new jobs and so contribute to their nation's export drive. This in itself will cause further trade friction between Japan and various western nations, so that the anti-whaling feeling which is—as we have already seen—partly based on trade friction may merely serve to make that friction worse.

Over recent years, the Japanese have responded to environmental arguments and tried to convince foreigners that there are enough whales to allow a modest and controlled exploitation thereof. But just when the evidence that they and others produce begins to support their argument so overwhelmingly that it becomes difficult to defend the moratorium on a scientific basis, westerners' tactics change. They suddenly cease to focus on the question of environment as biology, and shift instead to an "environmental" argument that concentrates solely on ethics and morality. The whale is made into something very "special"—one of those animals that is stronger, smarter, cuter, faster, more awesome, and so on *ad infinitum*, than others. Yet, when it comes to ethics of this kind, the Japanese are confused, for Buddhism has taught them that all life has the same value. This makes it difficult for them to grasp why it should be more morally wrong to kill a whale for food than to kill a cow or a pig for the same purpose. Indeed, most Japanese think it is worse to kill a domesticated animal than a wild one. "How can people kill an animal they have fed", is a comment often heard in Japanese whaling communities.

What we in fact have, then, are at least two distinct value systems, each firmly embedded in and conditioned by historical processes in the cultures where they prevail. It smacks of cultural imperialism, therefore—if not of racism—for one cultural group to impose its values on the other, especially when the latter is seen to be morally inferior to oneself. Japanese are—along with the Faroese, Icelanders and Norwegians—depicted as a barbaric and gruesome people who kill for pleasure. In their campaigns against whaling, animal rights groups carry on the dubious western tradition of forcing upon others the values which at any given moment happen to be in vogue at home. Indeed, in a number of respects, it would seem that the Japanese themselves form as anomalous a category in the context of advanced industrialized societies as whales do among mammals. Non-white orientals in a white western arena, more advanced techno-logically than their earlier developing competitors, the Japanese are still to some extent characterized by the latter as "primitive barbarians" in a civilized world. Like the whale that is betwixt and between because it is a "mammal off the land", the Japanese, too, are very much like "fish out of water" among western "first world" countries.

Given overall western attitudes towards their country, some Japanese are not convinced that the squabbles over whaling are really ethical questions at all. Rather, they see the whaling issue in a broader perspective in which trade friction and defence questions also play an important part. They believe that whaling has become a convenient symbol for those who try to create anti-Japanese feelings in western countries where many people already blame the Japanese for many of the economic problems found in the West.

Whether the whaling issue is motivated by ethical considerations or by growing anti-Japanese sentiments fuelled by the country's tremendous economic success—there is probably a little of both—the Japanese see the foreigners' attack on whaling as an attack on their culture in general. We have seen in Chapter 2 how Taiji—and to a lesser extent also Ayukawa and Arikawa—has made use of whale symbols to foster a strong community identity—itself seen as an important strategy for survival. But whales and whaling have in recent years become symbols to a large number of Japanese, and not just to those involved in the whaling industry. Perhaps not surprisingly, therefore, a considerable number of advertisements now include whales and dolphins in their images. Whales and whaling have become a metaphor for a greater issue, so far as the Japanese are concerned.

Since the end of the Pacific War, the Japanese have had some trouble in fixing on national symbols acceptable to everybody (both at home and abroad). The use of the national flag and hymn is still controversial in some circles, as was the continued presence on the "chrysanthemum throne" of the Emperor Shōwa until 1989. Moreover, the Constitution prohibits public sponsorship of religious institutions.[16] In this vacuum of national symbols whale meat has provided a particularly powerful image. As most Japanese are fond of whale meat and believe that the whales saved them from starvation after the war, there is little disagreement about the value of such meat. Moreover, since few other peoples eat whale meat, this habit also sets the Japanese apart from others. The Japanese thus become unique,[17]

---

[16] The flag, hymn, head of State, and Christianity are all important national symbols in most western countries, and in particularly the Americans make use of their flag and the Bible in—what are for many non-Americans—the most surprising circumstances.

[17] That the Japanese are not alone in eating whale meat—the Eskimos, Icelanders, Faroese, and Norwegians, for example, are also fond of meat—is simply ignored in this context. As is the fact that in countries like the UK, whale meat was widely

and the whaling issue serves to strengthen the much cherished Japanese myths about their identity (*nihonjinron*), which itself helps fuel one form of Japanese nationalism. Certainly, people have been using the whaling issue for that purpose in Japan, as they have done so for a different kind of nationalism in the United States. This is, perhaps, the most serious outcome of the anti-whaling campaigns and of the moratorium on commercial whaling.

Whaling has thus become a political question in Japan, too, but the whalers suffer the consequences of trying to participate in a game in which the rules are always being redefined by the opposition. It would seem to be time, therefore, for the referee to reinstate the goal posts and call the players to order, before the game degenerates further. Let us end this book with the words of a Japanese whaler:

> "We have worked hard to feed our children and our nation. We have taken better care of the whale carcass than any other whaler on Earth. We have prayed for their souls so that they can be reborn in Paradise, just as we do for our own father and mother. Why should we now suffer? If there were no more whales, we could accept it. But why should we suffer for political reasons? Why should *we* suffer as a result of other countries wanting to solve *their* problems?"

---

eaten during the Second World War.

Appendices

Appendix 1:
Annual Catches of Large Whales by Station, 1935-40

Station	1935	1936	1937	1938	1939	1940	Total
Satto	-	18	1	-	-	-	19
Shiberu	-	15	67	63	59	38	242
Shana	184	326	452	558	512	410	2,442
Hitokappu	135	91	73	98	71	69	537
Shakotan	131	28	74	86	20	23	362
Abashiri	2	-	-	-	-	30	32
Kiritappu	84	64	70	30	30	28	306
Akkeshi	146	26	77	65	82	98	494
Naiho	2	-	2	-	-	-	4
Kamaishi	32	-	-	-	5	38	75
Ayukawa	664	897	730	654	1,083	801	4,849
Taiji	40	-	-	-	-	-	40
Ōshima	48	47	73	91	70	103	432
Kushimoto	14	-	-	-	-	-	14
Tonoura	-	-	14	8	1	9	32
Yobuko	18	8	11	11	8	9	65
Arikawa	2	-	-	-	-	-	2
Koshikijima	-	-	1	-	-	-	1
Okochi	4	5	2	-	5	3	19
Saishuto	15	29	59	17	10	8	138
Urusan	64	76	82	116	30	90	458
Daikokusanto	52	20	72	39	30	3	216
Daiseito	9	-	-	-	-	3	12
Kaiyoto	-	-	-	-	4	11	15
Kunetsu	16	-	-	-	-	-	16
Dainbanratsu	33	20	22	17	9	10	111
Ogasawara	76	134	168	126	207	244	955
Total	1,771	1,814	2,050	1,979	2,298	2,035	11,888

Source: Terry 1950:46-47.

Appendix 2:
Production of Pelagic Whaling, LTCW and STCW

| Year | Pelagic | | Coastal | | Total |
	Antarctic	North Pacific	LTCW	STCW	(1,000 tons)
1932	0.0	0.0	22.5	not available	
1933	0.0	0.0	23.1		
1934	0.0	0.0	25.4		
1935	2.0	0.0	27.7		
1936	7.6	0.0	28.5		
1937	26.4	0.0	29.3		
1938	66.2	0.0	26.6		
1939	85.1	0.0	26.1		
1940	101.2	6.1	24.4		
1941	120.1	7.6	30.7		
1942	0.0	0.0	19.1		
1943	0.0	0.0	25.7		
1944	0.0	0.0	34.8		
1945	0.0	0.0	9.3		
1946	0.0	0.0	23.8		
1947	34.5	0.0	28.6		
1948	45.4	0.0	26.8		
1949	55.0	0.0	25.0		
1950	68.1	0.0	22.6		
1951	58.6	0.0	26.4	2.8	87.8
1952	71.0	6.5	28.8	3.4	109.7
1953	59.5	12.6	25.4	3.0	100.5
1954	73.9	34.0	25.9	2.6	136.4
1955	102.8	31.8	26.9	2.9	164.4
1956	115.2	42.1	33.4	3.5	194.2
1957	154.5	44.9	28.3	2.4	230.1
1958	196.4	45.3	33.3	2.9	277.9
1959	206.6	50.4	33.8	2.1	292.3
1960	213.8	52.9	27.9	1.9	296.5
1961	253.7	50.6	26.1	2.1	332.5
1962	309.5	61.4	25.3	1.9	398.1
1963	283.6	57.3	24.1	2.0	367.0
1964	274.9	61.5	23.6	2.5	362.5
1965	270.5	70.8	21.0	2.7	365.0
1966	170.6	83.7	23.3	2.3	279.9
1967	142.1	91.4	28.5	1.8	263.7
1968	125.9	87.9	39.5	1.6	254.9
1969	105.4	78.6	38.4	1.6	223.8
1970	110.5	81.0	34.6	1.4	227.6

Japanese Whaling

Year	Pelagic Antarctic	Pelagic North Pacific	Coastal LTCW	Coastal STCW	Total (1,000 tons)
1971	122.1	66.8	33.1	1.3	223.3
1972	116.0	50.6	26.8	1.1	194.5
1973	95.9	44.3	22.5	1.2	163.9
1974	80.9	41.8	21.9	0.9	145.5
1975	76.2	29.1	20.6	1.0	127.0
1976	41.1	13.1	22.4	0.6	77.3
1977	38.2	13.1	19.7	0.7	71.7
1978	17.6	6.1	17.5	0.9	42.2
1979	14.8	2.0	13.2	0.8	30.8
1980	16.4	0.0	15.1	0.6	32.1
1981	15.0	0.0	12.9	0.7	28.6
1982	17.9	0.0	9.0	0.7	27.6
1983	16.7	0.0	8.9	0.7	26.3
1984	15.6	0.0	8.5	1.1	25.2
1985	10.2	0.0	7.5	0.9	18.6
1986	10.6	0.0	5.5	0.9	17.0
1987	10.6	0.0	5.7	0.9	17.2
1988	0.0	0.0	0.0	0.5	0.5
1989	0.0	0.0	0.0	0.4	0.4

Sources: Tatō 1985:156-157, 164; le Grand et al., n.d.; Institute of Cetacean Research.

Appendix 3: The Structure of Net Groups

A: Sea Operations

Type of boat	Katsumoto 1802		Ukushima 1817		Ogawajima 1773	
	Boats	Crew	Boats	Crew	Boats	Crew
Chaser boat (*seko-bune*)	20	260	12	156	16	208
Net boats (*sōkai-sen*)	12	120	6	60	6	?
Assisting boats (*amitsuke-bune*)	12	144	6	72	6	?
Towing boats (*mossō-bune*)	8	96	4	48	4	?
Other boats			3	18	2	?
harpooners (*hazashi*)		40		25		incl.
Other crew members		10		1		3
No. of boats and persons	52	670	31	280	34	400?

B: A Shore Station in Kyūshū

Type of person employed	large shed	small shed	sinew shed	bone shed
managers (*shikai' nin*)	3	2	1	1
accountants (*kanjō-kata*)	2			
scribes (*chōmen-yaku*)		1		
intendents (*ko-bettō*)	2	1		
deputies (*mokudai*)	c.20	8	2	2
flensers (*uo-kiri*)	10	5		
cutters (*naka-kiri*)	6			
in charge of kettles (*kamakake*)	6			
"handy man" (*oimawashi*)	6	2		1
guards (*ban' in*)	3			
cooks (*meshi-taki*)	4	1	1	1
net-makers (*ami-daiku*)	2			
boat-builders (*funa-daiku*)	2			
smith (*kaji*)	1			
cooper (*okeya*)	1			
prepare sinew (*suji-koshirae*)	12			
fetching water (*mizukumi*)			1	
oil-collectors (*abura-tori*)				2
bone-crushers (*hone kezuri*)				1
Total seasonal/regular employees	c.80	20	5	8
day-labourers employed when whales were caught	100 -200	50 -70	few	50- 100

Sources: Hidemura (1952b:68, 73) and *Hizen-no-kuni sanbutsu zukō* (Vol.4) published in *Genkai no kujira tori*, Saga: Saga kenritsu hakubutsukan, 1980.

Appendix 4A:
The Homes of Whalers Employed by Nissui for the Antarctic Expedition of *Tōnan-maru* in 1965/66

Place of residence (prefecture)	Factory ship *Tōnan-maru* Crew	Workers	Catcher boats	Refrigerator ships (2) Crew	Workers	South Georgia	Tot
Kagoshima	1		2	3			6
Miyazaki			3				3
Kumamoto	3						3
Ōita			2	4			6
Nagasaki (*)	5	52	24	2	50	22	155
Saga	1		2	2			5
Fukuoka	5	3	9	2	2	1	22
Ehime	5	5	5	1			16
Tokushima	2	1	2				5
Kagawa	2	6	1			9	32
Kōchi	2		3	4	1		10
Yamaguchi	4	17	11	2	13	10	57
Hiroshima	15	10	13	14	14	4	70
Okayama	3		1	2	1		7
Shimane	3		6	5	1		15
Tottori	1						1
Hyōgo	6	2	25	5	3		41
Ōsaka	9	6	6	5	1	1	28
Kyōto	1	1	2	2		1	7
Nara	2						2
Wakayama			3	1			4
Mie	1		1				2
Fukui	8			3	1		12
Ishikawa	24	7	9	22	23	2	87
Tōyama	1	2	1	1	3	1	9
Niigata		1	8	1			10
Shizuoka			4	1			5
Kanagawa	13		6	12		1	32
Yamanashi			1				1
Saitama	1			3			4
Gumma			1				1
Tōkyō	9		4	6			19
Chiba	3	4	9	4	4		24
Ibaragi			7				7
Tochigi			1				1
Fukushima			2				2

Place of residence (prefecture)	Factory ship *Tōnan-maru* Crew	Workers	Catcher boats	Refrigerator ships (2) Crew	Workers	South Georgia	Tot
Yamagata			1				1
Miyagi	5	8	32	8	6	7	66
Iwate	1	3	5	1	13	4	27
Aomori	1	28	2		50	19	100
Akita		11			61	10	82
Hokkaidō	1	5	2	5	47	1	61
No. of pers.	137	167	215	128	309	93	1,049

Note:

*: Of the 52 workers from Nagasaki on *Tōnan-maru*, 38 were from Arikawa, six from Shinuonome, two from Ukushima and six from elsewhere in the prefecture. Of the 24 crew members on catcher boats, 13 were from Ukushima, four from Arikawa and seven from elsewhere. Of the 50 who worked on refrigerator ships, 44 persons came Arikawa, three from Kami-Gotō Town, one from Ukushima and one from elsewhere. 16 persons from Arikawa, three from Shinuonome and three from elsewhere in Nagasaki worked at South Georgia.

Source: Nihon Suisan (1966:2-13).

Appendix 4A is not meant to give a correct picture of the spatial distribution of the whalers in Japan, as the table is based on one Nissui expedition to the Antarctic only. Most conspiciously, whalers from Taiji were not employed by this company and relatively few, too, were employed from Ayukawa. However, the table indicates that the residences of the crew members on the factory and refrigerator ships were more widely distributed than those of the "workers", i.e. those handling the whales. The crew members were sailors, and typically many of them lived in cities, whereas the "workers" lived in the countryside. The crews on the catcher boats were also recruited from a wide area.

Appendix 4B:
Place of residence of the Nihon Hogei employees when they were laid off in 1987

Province	Town/City	Worked on catcher boats	Worked on flensing station
Miyagi	Oshika - Ayukawa	5 (a)	13
	Kugunari		3
	Tomari	1	
	Unknown hamlet		1 (e)
	Ishinomaki City	7	4
	Onagawa Town	1 (b)	
	Ogatsu	1	
	Yamoto	6	2 (e)
	Kogota	1	
	Naruse		1 (e)
	Shiogama City		1 (e)
	Sendai City	1 (c)	
Aomori	Hachinohe City	1	
	Oma		2
Kanagawa	Yokosuka City	1	
Fukui	Fukui City	1	
Wakayama	Taiji Town	13	
	Katsuura Town		1 (e)
Yamaguchi	Shimonoseki City	1 (d)	
	Ube City	1	
Fukuoka	Fukuoka City	1	
Nagasaki	Arikawa Town	1	
	Sasebo City	1	
	TOTAL	44	28

Notes:
a: one born in Hirado (Nagasaki Prefecture)
b: born in Niiyama, Oshika Town
c: born in Taiji
d: born in Kugunari, Oshika Town
e: born in Ayukawa, Oshika Town

Appendix 4C:
The places of birth for the present STCW crews

| | Place of birth | | |
Boat	Gunner	Captain	Crew
Yasu-maru No.2 (Abashiri)	Aomori Pref.	Akita Pref.	Abashiri 2 Hokkaidō 3 Arikawa 1
Takashima-maru No.8 (Abashiri)	Ishinomaki	Kagawa Pref.	Abashiri 3 Hokkaidō 2 Iwate Pref. 1
Kōei-maru No.75 (Ayukawa)	Ishinomaki	Ayukawa	Oshika Town 4 Ishinomaki 1 Abashiri 1
Taishō-maru No.2 (Ayukawa)	Ajishima, Oshika Town	Ayukawa	Oshika Town 1 Ishinomaki 2 Miyagi Pref. 2
Taishō-maru (Ayukawa)	Niigata Pref.	Yamaguchi Pref.	Oshika Town 3 Miyagi Pref. 2 Hokkaidō 1
Sumitomo-maru No.21 (Wadaura)	Miyagi Pref.	Ayukawa	Oshika Town 5
Katsu-maru (Taiji)	Taiji	Taiji	Taiji 4
Seishin-maru (Taiji)	Taiji	Taiji	Taiji 6

Appendix 5:
The Geographical Background of Household Heads in Ayukawa

Prefecture	County	Township	Section	Number	%
Miyagi	Oshika	Oshika	Ayukawa	378	28.2
"		"	others	153	11.4 (a)
"	"	others, incl Ishinomaki		181	13.5 (b)
"		Senenmiyagi incl. Sendai, Shiogama		99	7.4
"		Monō		99	7.4
"		Shida		25	1.9
"		others		120	9.0
Hokkaidō				11	0.8
Aomori				8	0.6
Akita				10	0.7
Iwate				41	3.1
Niigata				9	0.7
Yamagata				23	1.7
Fukushima				21	1.6
Tōkyō				25	1.9
Ishikawa				8	0.6
Wakayama (*)				8	0.6
Kōchi (*)				10	0.7
Nagasaki (*)				18	1.3
Yamaguchi (*)				16	5.7
Others				77	5.7
TOTAL				1,340	100.0

Notes:
a: These household heads came from Kugunari 33, Niiyama 22, Ōhara 20, Kobuchi 17, Kyūbun 12, Tomari 12, Ajishima 10,
b: Among these 76 came from Ishinomaki and 25 from Watanoha, now a part of Ishinomaki City.
*: Old whaling centres.

Source: Oshika-chō 1988:129-130.

Bibliography

Akashi Documents. Tokugawa period documents kept at Fukuoka prefectural library.

Akimichi T., P.J. Asquith, H. Befu, T.C. Bestor, S.R. Braund, M.M.R. Freeman, H. Hardacre, M. Iwasaki, A. Kalland, L. Manderson, B.D. Moeran and J. Takahashi
1988 *Small-Type Coastal Whaling in Japan*, Edmonton: Boreal Institute for Northern Studies, Occasional Paper No.27.

Andersen, Raoul
1972 "Hunt and deceive: Information management in Newfoundland deep-sea trawler fishing". In R. Andersen and C. Wadel (eds.) *North Atlantic Fishermen*. St.John's: Memorial University of Newfoundland.

Andresen, Steinar
1989 "Science and politics in the international management of whales". *Marine Policy*, 13(2):99-117.

Arikawa-chō
1982 *Arikawa no ayumi*. Arikawa: Arikawa-chō.

Aron, William
1988 "The commons revisited: thoughts on marine mammal management". *Coastal Management*, 16(2):99-110.

Barstow, Robbins
1989 "Beyond whale species survival: Peaceful coexistence and mutual enrichment as a basis for human/cetacean relations". *Sonar*, No.2:10-13.

Barth, Fredrik
1956 "Ecological relationships of ethnic groups in Swat, Northern Pakistan". *American Anthropologist*, 58:1079-89.

1966 *Models of Social Organization*. London: Royal Anthropological Institute of Great Britain & Ireland, Occasional Paper No.23.

Beardsley, Richard K., John W. Hall and Robert W. Ward
1969 *Village Japan*. Chicago: University of Chicago Press.

Befu, Harumi
1980 "Political Ecology of Fishing in Japan: Techno-environmental Impact of Industrialization in the Inland Sea". *Research in Economic Anthropology*, 3:323-347.

Bestor, Theodore C.
1985 "Tradition and Japananese Social Organization: Institutional Development in a Tokyo Neighborhood." *Ethnology*, 24(2):121-135.

1989a *Neighborhood Tokyo*. Stanford: Stanford University Press.

1989b "Socio Economic Implications of a Zero Catch Limit on Distribution Channels and Related Activities in Hokkaido and Miyagi Prefecture, Japan". Report IWC/41/SE1.

Byron, Reginald
1980 "Skippers and strategies: leadership and innovation in Shetland fishing crews". *Human Organization*, 39(3):227-232.

Christie, Francis T. and Anthony Scott
1965 *The Common Wealth of Ocean Fisheries*. London: John Hopkins Press.

Clark, Rodney
1979 *The Japanese Company*. New Haven: Yale University Press.

Cole, Robert E.
1971 *Japanese Blue Collar: the Changing Tradition*. Berkeley and Los Angeles: University of California Press.

Corrigan, Patricia
1991 *Where the Whales Are: Your Guide to Whale-Watching Trips in North America*. Chester, Conn.: The Globe Pequot Press.

Dore, Ronald
1958 *City Life in Japan*. Berkeley and Los Angeles: University of California Press.

1959 *Land Reform in Japan*. London: Oxford University Press.

1973 *British Factory - Japanese Factory*. Berkeley and Los Angeles: University of California Press.

1978 *Shinohata: Portrait of a Japanese Village*. London: Allen Lane.

Douglas, Mary
1966 *Purity and Danger: An Analysis of the Concepts of Pollution and Taboo.* London: Routledge and Kegan-Paul.

Embree, John F.
1939 *Suye Mura: A Japanese Village.* Chicago: University of Chicago Press.

Eyerman, Ron and Andrew Jamison
1989 "Environmental knowledge as an organizational weapon: the case of Greenpeace". *Social Science Information,* 28:99-119.

Ferrari, Maxime
1983 "Of whales and politics". *Ambio,* 12(6):347-348.

Forman, Shepard
1967 "Cognition and the catch: the location of fishing spots in a Brazilian coastal village". *Ethnology,* 6(4):417-426.

Fox, John
1991 "The business of Greenpeace". *The Financial Post* (Toronto), 7 January.

Freeman, Milton M.R.
1990 "A commentary on political issues with regard to contemporary whaling". *North Atlantic Studies,* 2(1-2):106-116.

Fujimoto Takashi
1964 "Bakumatsu Saikai hogeigyō no shikin kōsei: Ikitsukishima Masutomi no baai". In *Sōritsu 30-nen Fukuoka Daigaku kinen ronbunshū: shōgaku-hen,* pp. 251-287.

1967 "Geiyu no ryūtsu to chihō shijō no keisei", *Kyūshū bunkashi kenkyūjo kiyō,* 12:125-154.

Fujimoto Takashi, K. Kubota, and K. Hara
1984 "Arikawa kujira-gumi shiki hōtei", I and II. *Fukuoka Daigaku shōgaku ronsō,* 28(3):225-260 and 28(4):633-685.

Fukumoto Kazuo
1978 *Nihon hogeishi-wa.* Tōkyō: Hōsei University Press.

Fukutake Tadashi
1949 *Nihon nōson no shakaiteki seikaku.* Tōkyō: Tōkyō University Press.

1956 "Gendai Nihon ni okeru sonraku kyōdōtai sonzai keitai", in *Sonraku kyōdōtai no kōzō bunseki.* Sonraku Shakai Kenkyūkai Nenpo No.3:1-20.

Fukutake Tadashi
1982 *The Japanese Social Structure: Its Evolution in the Modern Century.* Tōkyō: University of Tōkyō Press.

Gatewood, J.B.
1984 "Corporation, competition, and synergy: informationsharing groups among southeast Alaska salmon seiners". *American Ethnologist,* 11(2):350-367.

Government of Japan
1989 "Report to the Working Group on Socio-economic Implications of a Zero Catch Limit". Report IWC/41/21.

le Grand, C., T. Tsukaguchi-le Grand, and G.O. Totten
n.d. *Japanese Whaling - An Industry in Decline.* Unpublished manuscript.

Gulland, John
1988 "The end of whaling?". *New Scientist,* 29 October, pp.42-47.

Hayami Tadashi
1981 *Labor Relations in Japan Today.* Tōkyō: Kodansha.

Henke, Janice S.
1985 *Seal Wars. An American View.* St.Johns: Breakwater Books, Ltd.

Herscovici, Alan
1985 *Second Nature - The Animal-Rights Controversy.* Montreal: CBC Enterprises.

Hidemura Senzō
1952a "Tokugawa-ki Kyūshū ni okeru hogeigyō no rōdo kankei, (1)". *Kyūdai keizaigaku kenkyū,* 18(1):57-85.

1952b "Tokugawa-ki Kyūshū ni okeru hogeigyō no rōdo kankei, (2)". *Kyūdai keizaigaku kenkyū,* 18(2):67-108.

Hidemura Senzō and Fujimoto Takashi
1978 "Saikai hogeigyō". In *Edo-jidai zushi - Kyūshū I,* Tōkyō: Chikuma shobō, pp.160-169.

Hoel, Alf Håkon
1986 *The International Whaling Commission, 1972-1984: New Members, New Concerns.* Lysaker, Norway: The Fridtjof Nansen Institute, 2nd edition.

Holt, Sidney
1985 "Whale mining, whale saving". *Marine Policy,* 9(3):192-213.

The Humane Society of the United States
1987 "Small-Type Commercial Whaling in Japan". Report, Washington D.C.

Hunter, Robert,
1979 *Warriors of the Rainbow: A Chronical of the Greenpeace Movement.* New York: Rinehart and Winston.

Isoda S.
1951 "Sonraku kōzō no futatsu no katachi", *Hōshakaigaku,* 1:50-64.

Iwasaki Masami
1987 "Cultural Significance of Whaling in a Whaling Community in Abashiri". M.A.thesis, University of Alberta, Department of Anthropology.

Jinoshima-ura shōya kiroku. Unpublished document kept by a village headman. A copy is available at Nishi Nihon Bunka Kyōkai, Fukuoka City.

Japan Whaling Association
1954 *Japanese Whaling Industry,* Tōkyō: Japan Whaling Association.

Johnson, Chalmers
1982 *MITI and the Japanese Miracle.* Stanford: Standford University Press.

Kalland, Arne
1981 *Shingū - A Study of a Japanese Fishing Community.* SIAS Monograph Series No.44, London: Curzon Press.

1986 "Pre-modern whaling in Northern Kyūshū". In Erich Pauer (ed.) *Silkworms, Oil and Chips ...* Bonner Zeitschrift für Japanologie, Vol. 8:29-50.

1988 *Fishing Villages in Tokugawa Japan - The Case of Fukuoka Domain.* Unpublished thesis, University of Oslo.

1989 "Arikawa and the impact of a declining whaling industry". *NIAS Report* (Copenhagen), 1:94-138.

1990 "Sea tenure and the Japanese experience: resource management in coastal fisheries". In E. Ben-Ari, B. Moeran and J. Valentine (eds.): *Unwrapping Japan.* Manchester: Manchester University Press, pp.188-204.

1991a "Management by totemization: Whale symbolism and the anti-whaling campaign". Paper presented at the 2nd Annual International Association for the Study of Common Property Conference, Winnipeg, 26-29 September.

1991b "Whose whale is that? Diverting the commodity path". Paper read at the seminar *Com-oddities*, SOAS, Dept. og Anthropology, 11 November.

Klinowska, Margaret
1988 "Are cetaceans especially smart?". *New Scientist,* 29 October, pp.46-47.

Kumano Taiji-ura Hogei-shi Hensan I'inkai
1965 *Kumano no Taiji - kujira ni idomu machi.* Tōkyō: Heibonsha.

1969 *Kumano Taiji-ura hogeishi.* Tōkyō: Heibonsha.

Lynge, Finn
1990 *Kampen om de vilde dyr - en artisk vinkling.* Copenhagen: Akademisk Forlag.

Löfgren, Orvar
1979 "Marine ecotypes in preindustrial Sweden: A comparative discussion of Swedish peasant fishermen". In Raoul Andersen (ed.) *North Atlantic Maritime Culture.* The Hague: Mouton, pp.83-110.

Maeda Keijirō and Teraoka Yoshiro
1952 *Hogei.* Tōkyō: Nihon Hogei Kyōkai.

Malinowski, Bronislaw
1922 *Argonauts of the Western Pacific.* London: Routledge and Kegan Paul.

Manning, Laura
1989 "Marine mammals and fisheries conflicts: A philosophical dispute". *Ocean & Shoreline Management,* 12(3):217-232.

Marcus, Grel
1989 "We are the World?". In Angela McRobbie (ed.) *Zoot Suits and Second-Hand Dresses: Anthology of Fashion and Music.* London: Macmillan, pp.276-282.

Marsh, Robert M. and Hiroshi Mannari
1976 *Modernization and the Japanese Factory.* Princeton: Princeton University Press.

Martinez, Dolores
1990 "Tourism and the *ama*: the search for the real Japan". In E. Ben-Ari, B. Moeran and J. Valentine (eds.): *Unwrapping Japan.* Manchester: Manchester University Press, pp.97-116.

Moeran, Brian
1984 *Lost Innocence: folk craft potters of Onta, Japan.* Berkeley and Los Angeles: University of California Press.

1985 *Okubo Dairy: portrait of a Japanese Valley.* Stanford: Stanford University Press.

Möbius, K.
1893 "Über den Fang und die Verwerthung der Walfisches in Japan". In *Sitzungsberichte der Königlich Preussischen Akademie der Wissenschaften zu Berlin,* pp.1053-1073.

Mønnesland, J., S. Johansen, S. Eikeland, and K. Hanssen
1990 "Whaling in Norwegian waters in the 1980'ies: The economic and social aspects of the whaling industry, and the effects of its termination". Oslo: Norwegian Institute for Urban and Regional Research, *NIBR-report*, 1990-14.

Nagasaki Fukuzō
1989 "The Facts, 'Facts' and Fiction of Scientific Whaling". *Science & Technology in Japan*, 8(31):36-41.

Naitō Kanji
1970 "Inheritance practices on a Catholic island: youngest-son inheritance (ultimogeniture) on Kuroshima, Nagasaki Prefecture". *Social Compass*, 17:21-36.

Nakane Chie
1967 *Kinship and Economic Organization in Rural Japan*. London: The Athlone Press.

1970 *Japanese Society*. Middlesex: Penguin Books.

Nakayama Tomonori
1987 *Furusato no ayumi*. Arikawa: Arikawa-chō yōiku i'inkai.

Nihon Suisan
1966 *Dai 20-ji nangei*. Tōkyō: Nihon Suisan.

Nishi Nihon Bunka Kyōkai (eds.)
1988 *Fukuoka-kenshi: Kinsei shriryōhen - Fukuoka-han goyōcho, 1*. Fukuoka: Fukuoka Prefecture.

Noma Yoshio
1973 *Genkai no shimajima*. Tōkyō: Keiyūsha.

Norbeck, Edward.
1954 *Takashima: A Japanese Fishing Community*. Salt Lake City: University of Utah Press.

Odner, Knut
1978 "Encapsulation in a capitalist economy and traditional norms: Contributing factors to ecological imbalance among the Kenyan Masai". *Norsk Geografisk Tidsskrift*, 32:27-40.

Olsen, Bjørnar
1982 "Mehamn 1903 - fiskeopprør mot kvalfangstnæring". *Ottar*, 138(5):26-30.

Ōmori Ichinosuke
n.d.   "Ayukawa-ko no hogeishi". To be published in Volume 2 of *Oshika-chōshi*. Ayukawa: Oshika-chō yakuba.

Oshika-chō
1988   *Oshika-chōshi*, Vol.1. Ayukawa: Oshika-chō yakuba.

Pearse, Fred
1991   *Green Warriors: The People and the Politics behind the Environmental Revolution*. London: The Bodley Head.

Plath, David
1964   "Where the Family of God Is the Family: The Role of the Dead in Japanese Households". *American Anthropologist*, 66(2):300-317.

Plummer, Katrine
1984   *The Shogun's Reluctant Ambassadors - Sea Drifters*. Tōkyō: Lotus Press.

Prescott, John H.
1981   "Clever Hans: Training the trainers. On the potential for misinterpretating the results of dolphin research". *Annals of the New York Academy of Science*, 364.:103-136.

Pryor, Karen
1981   "Why Porpoise Trainers Are Not Dolphin Lovers: Real and False Communication in the Operant Setting". *Annals of the New York Academy of Science*, 364.:137-143.

Rohlen, Thomas P.
1974   *For Harmony and Strenth*. Berkeley and Los Angeles: University of California Press.

1983   *Japan's High Schools*. Berkeley and Los Angeles: University of California Press.

Ruddle, Kenneth
1987   *Administration and conflict management in Japanese coastal fisheries*. FAO Fisheries Technical Paper, No.273.

Sahlins, Marshall
1972   *Stone Age Economics*. London: Tavistock Publications.

Schwartz, Ulrich
1991   "Geldmachine Greenpeace". *Der Spiegel*, 45(38), 16 September, pp.84-105.

Schweitzer, Albert
1950 *The Animal World of Albert Schweitzer: Jungle Insight into Reverence for Life*. (Charles R. Joy, trans. and ed.). Boston: Beacon Press.

Shimada, Bell M.
1947 *Japanese Whaling in the Bonin Island Area*. SCAP General Headquarter, Tōkyō: Natural Resources Section Report No. 73.

Shimpo Mitsuru
1976 *Three Decades in Shiwa: Economic Development and Social Change in a Japanese Farming Community*. Vancouver: University of British Columbia Press.

Smith, Robert J.
1974 *Ancestor Worship in Contemporary Japan*. Stanford: Stanford University Press.

1978 *Kurusu: the Price of Progress in a Japanese Village 1951-75*. London: Dawson.

Spencer, L., J. Bollwerk and R.C. Morais
1991 "The not so peaceful world of Greenpeace". *Forbes Magazine*, 11 November, pp.175-180.

Statler, Oliver
1984 *Japanese Pilgrimage*. London: Picador.

Stokke, Olav Schram
1991 "Transnational fishing: Japan's changing strategy". *Marine Policy*, July, pp.231-243.

Stuster, Jack
1978 "Where 'mabel' may mean 'sea bass'". *Natural History*, 87(9):65-71.

Suenobu Genzō, Imabayashi Matsumi & Kubota Takeshi
1980 *Wajiro no ayumi*. Fukuoka City: no publisher

Taiji Gorōsuke
1982 *Kumano Taiji-ura hogei no hanashi*. Wakayama: Miyai Heiandō.

Taiyō Gyogyō
1984 *Maru-ha OB-kai kai'in meibo*. Tōkyō: Taiyō Gyogyō.

Takahashi Junichi
1987 "Hogei no machi no chōmin aidenteitī to shimboru no shiyō ni tsuite". *Minzokugaku kenkyū*, 52(2):158-167.

Takahashi Junichi
1988 *Women's Tales of Whaling: Life Stories of 11 Japanese Women who Lived with Whaling.* Tōkyō: Japan Whaling Association.

Takahashi Junichi, Arne Kalland, Brian Moeran and Theodore C. Bestor
1989 "Japanese Whaling Culture: Continuities and Diversities". *Maritime Anthropological Studies*, 2(2):105-133.

Takeno Yōko
1979 "Shiryō 'Maeme-Katsumoto kujira-gumi iezoku-kan'". *Fukuoka Daigaku shōgaku ronsō*, 24(1):105-165.

Tanaka Shogo
1987 *Kujira monogatari: nanpyō-yō o kaketa hōshu.* Tōkyō: Shibata Shoten.

Tatō Seitoku
1985 *Hogei no rekishi to shiryō.* Tōkyō: Suisansha.

Taylor, Bron
1991 "The Religion and Politics of Earth First!". *The Ecologist*, 21(6):258-266.

Terhune, Jack
1985 "Marine Survival". *Policy Options Politiques*, May, pp.24-26.

Terry, William M.
1950 *Japanese Whaling Industry Prior to 1946.* SCAP General Headquarters, Tōkyō: Natural Resources Section Report No.126.

Toby, Ronald
1984 *State Diplomacy in Early Modern Japan.* Princeton: Princeton University Press.

Tokumi Kōzō
1971 *Chōshū hogei-ko.* Shimonoseki: Nagato chihō shiryō kenkyūjo.

Tōyō Hogei
1910 *Honpo no Noruue-shiki hogeishi.* Tōkyō: Tōyō Hogei.

Tønnessen, Joh. N.
1967 *Den moderne hvalfangsts historie*, Vol.2. Sandefjord: Norges Hvalfangstforbund.

Tønnessen, Joh. N. and Arne Odd Johnsen
1982 *The History of Modern Whaling.* London: C.Hurst & Co.

US Marine Mammal Commission
1991 "Issues facing the International Whaling Commission and the U.S. regarding the resumption of commercial whaling and the future conservation of cetaceans". Memo.

Vogel, Ezra.
1963   *Japan's New Middle Class*. Berkeley and Los Angeles: University of California Press.

Wada-chō
1987   *Tōkei shiryō - Shōwa 62 nenpan*. Wada: Wada-chō.

Wenzel, George
1991   *Animal Rights, Human Right - Ecology, Economy and Ideology in the Canadian Arctic*. Toronto: University of Toronto Press.

Williams, Raymond
1973   *The Country and the City*. London: Chatto & Windus.

Yoshida Keiichi
1972   "Arikawa-chō hogeishi". In *Arikawa-chō kyōdōshi shiryōhen*. Arikawa: Arikawa-chō.

Zulaika, Joseba
1981   *Terranova: The Ethos and Luck of Deep-Sea Fishermen*. Philadelphia: ISHI.

Index